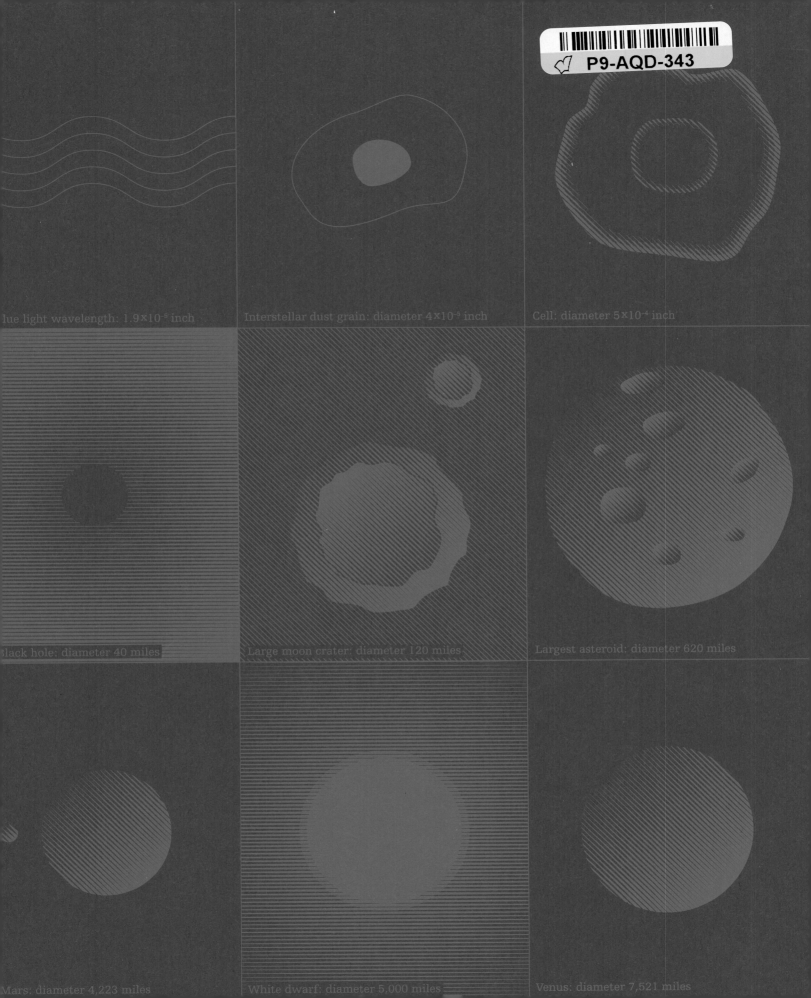

Blue light wavelength: 1.9×10^{-5} inch

Interstellar dust grain: diameter 4×10^{-5} inch

Cell: diameter 5×10^{-4} inch

Black hole: diameter 40 miles

Large moon crater: diameter 120 miles

Largest asteroid: diameter 620 miles

Mars: diameter 4,223 miles

White dwarf: diameter 5,000 miles

Venus: diameter 7,521 miles

MOONS AND RINGS

Other Publications:
THE NEW FACE OF WAR
HOW THINGS WORK
WINGS OF WAR
CREATIVE EVERYDAY COOKING
COLLECTOR'S LIBRARY OF THE UNKNOWN
CLASSICS OF WORLD WAR II
TIME-LIFE LIBRARY OF CURIOUS AND UNUSUAL FACTS
AMERICAN COUNTRY
THE THIRD REICH
THE TIME-LIFE GARDENER'S GUIDE
MYSTERIES OF THE UNKNOWN
TIME FRAME
FIX IT YOURSELF
FITNESS, HEALTH & NUTRITION
SUCCESSFUL PARENTING
HEALTHY HOME COOKING
UNDERSTANDING COMPUTERS
LIBRARY OF NATIONS
THE ENCHANTED WORLD
THE KODAK LIBRARY OF CREATIVE PHOTOGRAPHY
GREAT MEALS IN MINUTES
THE CIVIL WAR
PLANET EARTH
COLLECTOR'S LIBRARY OF THE CIVIL WAR
THE EPIC OF FLIGHT
THE GOOD COOK
WORLD WAR II
HOME REPAIR AND IMPROVEMENT
THE OLD WEST

This volume is one of a series that examines the universe in all its aspects, from its beginnings in the Big Bang to the promise of space exploration.

VOYAGE THROUGH THE UNIVERSE

MOONS AND RINGS

BY THE EDITORS OF TIME-LIFE BOOKS
ALEXANDRIA, VIRGINIA

CONTENTS

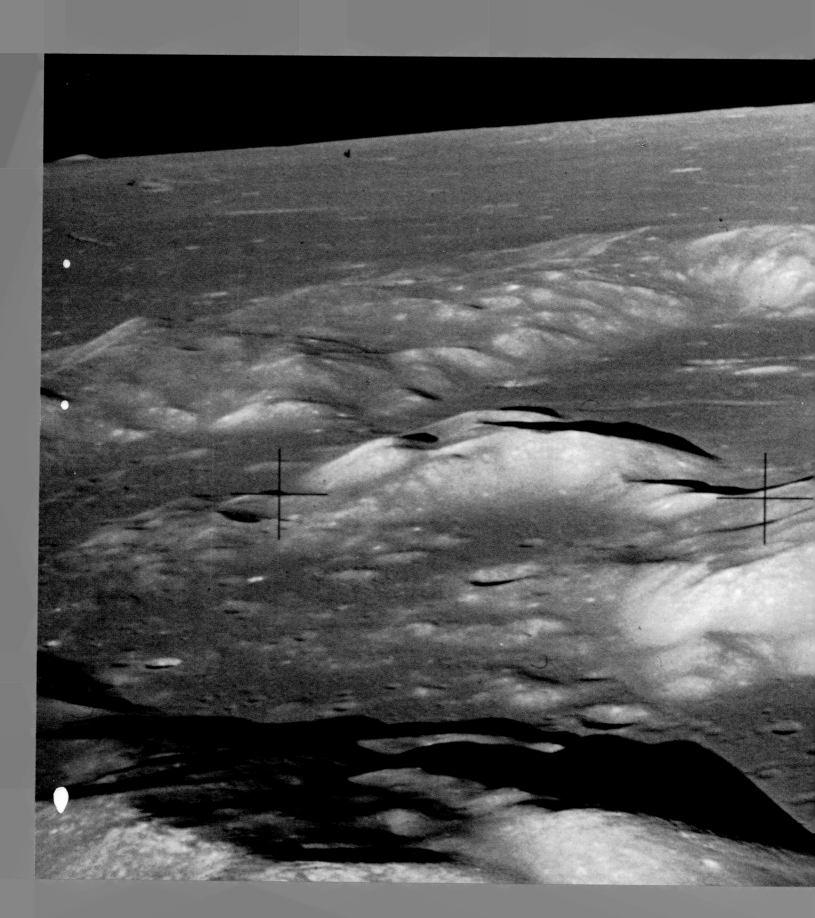

Framed by shadowed hills and the gleaming South Massif *(center),* the Taurus-Littrow valley *(foreground)* passes below the tiny command module of *Apollo 17* in a view shot from the just-separated lunar lander in December 1972. The lunar missions yielded crucial geologic evidence for scientists probing the Moon's origin.

J ames W. Christy had no inkling that he was on the verge of a major discovery. As the thirty-nine-year-old astronomer arrived for work at the U.S. Naval Observatory in Washington, D.C., on the morning of June 22, 1978, his mind was on a down-to-earth matter—the move he and his family would soon be making to a new home in the area. Christy would be doing most of the hauling himself, a dismal prospect in the sticky heat then blanketing the capital. Under the circumstances, the observatory's dark, air-conditioned measuring room, where Christy would spend the day poring over photographic plates of the planet Pluto, seemed a welcome refuge, and he turned to his task with an uncommon sense of relaxation.

Christy's objective was to chart the orbit of distant Pluto by measuring its position in reference to background stars, as shown in a sequence of photographs taken earlier that year through the sixty-one-inch reflecting telescope at the Naval Observatory's Flagstaff Station. At first glance, the results of the survey seemed disappointing. The image of Pluto on the plates was elongated, suggesting that the telescope's tracking device had erred slightly as it compensated for the Earth's rotation during each ninety-second exposure. But when Christy projected the apparently defective images onto a three-foot screen in the measuring room, he noticed something curious: The reference stars were not similarly elongated, as might be expected if the fault lay with the telescope at Flagstaff or some other distorting factor.

Intrigued, Christy put the plates under a microscope and examined Pluto more closely. In recent years, he had scrutinized thousands of photographs of binary stars, which lay so near each other that they often melded into one blurred image. Drawing on that experience now, the astronomer made an intuitive leap: The elongation of Pluto was not a defect at all, but the telltale sign of a second body in close proximity to the planet. It did not take him long to deduce the relationship of that body to Pluto. In a photograph taken April 13, he noticed, the planet was elongated to the south, while in another image produced a month later, Pluto bulged out to the north. The conclusion was inescapable—a moon was orbiting the ninth planet.

In one sense, Christy's serendipitous discovery of this distant body— dubbed Charon for the mythological boatman said to ferry the dead to Pluto's underworld—marked the end of an era. For more than three and a half centuries, earthbound observers had been extending the scope of their knowledge of moons and rings toward the outer reaches of the Solar System by refining their instruments and sharpening their analytical skills. The great Italian scientist, Galileo Galilei, had launched the effort in 1610, when he used a new device, the telescope, to spot four moons around Jupiter as well as curious appendages around Saturn that were ultimately identified as rings. A short time later, German astronomer Johannes Kepler speculated on philosophical grounds that Mars might have two moons, thus allowing for an orderly progression from Earth's solitary satellite to Jupiter's quartet. As it happened, Kepler's hunch turned out to be correct. But confirmation would not come until the late nineteenth century, by which time astronomers were developing photographic techniques that would lead to the identification of many small satellites orbiting Jupiter and the planets beyond in a pattern that vastly complicated the emerging portrait of the Solar System.

Christy's use of photoanalysis to spot Pluto's Charon carried this process of discovery to the known limits of the Sun's domain. But much was still to be learned about the territory lying between Earth and this distant outrider. Within a year of Christy's find, the twin American probes *Voyager 1* and *2* flew by Jupiter, providing views of that huge gaseous body and its satellites that constituted as great an advance for planetary science as Galileo's first telescopic forays. Over the next decade, knowledge of the satellite systems of the outer planets increased dramatically as the Voyager spacecraft continued on their separate paths, bringing Saturn, Uranus, and finally Neptune under close scrutiny. Thanks to the probes and to increasingly sensitive earthbound instruments, more than twenty new moons have been detected since the Voyager missions began in 1977, bringing the total in the Solar System to sixty-one and counting. In addition to finding a host of new satellites, scientists have also spotted rings around Jupiter, Neptune, and Uranus to complement those already known to exist around Saturn.

Nor have the gains been merely quantitative. As researchers evaluate the data, they are beginning to understand the forces that have yoked bodies large and small—including the fragments composing rings—into orbit around the planets. Astronomers now regard the astounding diversity of natural satellites, which range in scope from mere shards to bodies larger than Mercury, as compelling evidence that random impacts have substantially reshaped the Solar System since its formation some 4.6 billion years ago. But while collisions have been the wild card in the scheme, gravity has tended to impose order on satellite systems, guiding many moons into neat orbital patterns that seem to bear out the expectations of past astronomers who looked for harmony and precision in the universe.

No body better exemplifies the dual nature of satellite dynamics—tidy in design yet conditioned by calamity—than Earth's own moon. So regular in shape and orbit that it was long regarded as a flawless sphere and served as the basis for most early calendars, the Moon remains a model of consistency, an abiding presence that spins once on its axis for every revolution about

Earth so that it always presents the same face to the planet. Yet that constant face bears the marks of sporadic cosmic violence in its many craters. Indeed, after puzzling over the origins of the Moon for more than a century, many scientists have concluded that this serene beacon was born of an ancient cataclysm, perhaps the most devastating impact Earth has ever known.

THE ROAD TO LUNAR SCIENCE
Since the dawn of civilization, Earth's luminous moon has challenged the intellect and fired the imagination, inspiring both careful study and wild speculation. More than 4,000 years ago, the comings and goings of the Moon were meticulously recorded in the annals of Sumerian priests, who charted its progress across the heavens from the mud-brick towers of massive ziggurats. The Sumerian passion for order, which fostered an intricate network of canals and city-states in sunbaked Mesopotamia, also made for precise astronomical readings. Surviving cuneiform tablets indicate that Mesopotamian sky watchers devised accurate lunar calendars and could predict lunar eclipses. Yet they also worshiped the Moon as a god and read into its changes all manner of portents. However predictable, lunar eclipses were regarded as ominous events, and special priests were called on to exorcise the demons

AN ASSORTMENT OF MOONS

Closest to the Sun, Mercury and Venus claim no satellites, but the third planet, Earth, is circled by one of the largest moons in the Solar System, with a diameter of 2,160 miles. Mars, next out, has two companions—Phobos and Deimos—so tiny that they were not discovered until long after many of the major moons of the far planets had been found.

The four largest of Jupiter's sixteen known moons, known as the Galilean satellites, range in size from slightly smaller than the Moon to bigger than Mercury. The remainder—four very close to the planet and eight quite distant—are so diminutive that the brightest of them eluded detection for 282 years after Galileo's finds.

Tilted Saturn's eighteen moons constitute the biggest brood in the Solar System. Titan, 3,200 miles wide and first to be discovered, in 1655, dominates the group. By the late 1700s, six other major satellites of lesser brightness had been discerned. Of the later finds, the largest is Hyperion, at roughly 200 miles long and 125 miles wide.

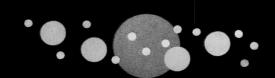
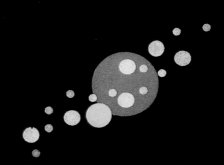

that had supposedly attacked the Moon god and obscured it from view.

A similar blend of shrewd reckonings and raw superstition characterized the Moon lore of other ancient peoples, including the Chinese and the Egyptians. With the rise of Greek philosophy in the sixth century BC, lunar facts began to take precedence over fables. But even the Greeks labored under the notion that the Moon, as a heavenly body, should be pure in form and perfect in orbit. The natural philosopher Anaxagoras, born around 500 BC, was banished from Athens for impiety after he suggested that the Moon was made of the same stuff as Earth, rather than a more exalted substance. And Aristarchus of Samos—who accomplished one of the great feats of ancient astronomy around 280 BC when he used a method of triangulation to calculate the relative distances from Earth of the Moon and the Sun—nonetheless assumed that the Moon follows an exact circular orbit around Earth rather than the slightly more elliptical path it in fact traces. Still, the Greek penchant for subjecting received wisdom to the test of reason laid a solid foundation for lunar science. When Aristotle sought to prove the assumption that Earth is a sphere, for instance, he offered an insight into the nature of lunar eclipses that had eluded the worshipful Sumerians. Earth's round shape, he concluded, is evident in the shadow cast across the Moon during an eclipse: "Since

Fifteen moons orbit Uranus's equator in a plane nearly perpendicular to Earth's orbital plane, describing concentric circles around upended Uranus. The planet is so distant that even its five major moons—led by Titania, nearly 1,000 miles across—were hard to spot; 300-mile-wide Miranda was not found until 1948. *Voyager 2* spied ten more moons in the mid-1980s.

Of the eight known satellites orbiting Neptune, only Triton, almost 1,700 miles across, was large enough to be detected before the 1900s. A second, much smaller moon—distant Nereid—was spotted through photographic analysis in 1949 by Gerard Kuiper, who used the same technique to find Uranus's Miranda. Six tiny orbs were spied by *Voyager 2* in 1989.

Pluto's sole satellite, 740-mile-wide Charon, is the most massive known moon relative to its primary; the two bodies orbit their common center of gravity, separated by just 12,000 miles. More than three billion miles away from Earth on average, the two were indistinguishable until James Christy identified Charon as a bulge on photographic images of Pluto in 1978.

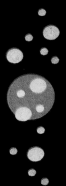

THE MOON'S MANY ASPECTS

Sky watchers have observed since ancient times that although the Moon shows a constant face to Earth, it changes shape nightly—now round and full, now the merest sliver, now vanishing altogether for an evening or two. On more dramatic occasions, the full moon is swallowed by an encroaching darkness; at other times, when the Moon is absent at night, it crosses the daytime sky to swallow the Sun. All of these phenomena derive from certain characteristics of the lunar orbit, as illustrated here.

No matter where the Moon is in its circuit around Earth, half of its surface is always illuminated by the Sun *(below)*. But the amount visible in the night sky over the planet varies in a regular cycle *(bottom)* that depends on the angle between the Sun, the Moon, and Earth. Eclipses can occur only when the trio lines up—which is also roughly the configuration that produces a new moon, when none of the Moon's sunlit side can be seen from Earth, or a full moon, when all of it can. As illustrated at right, only a slight inclination of the plane of the Moon's orbit with respect to the plane of Earth's orbit around the Sun prevents eclipses from being mundane monthly events—and preserves full moons for romantics on Earth.

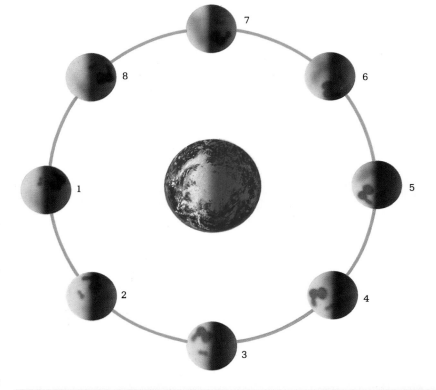

Spinning ever so slowly on its axis, the Moon completes one axial rotation in exactly the same time it takes to circle once around Earth. As depicted in the overhead view at right, the matched rates keep the same lunar hemisphere turned toward the planet at all times, so that observers see the same pattern of lunar craters. However, because of the shifting angle between the satellite and its primary with respect to the Sun, the amount of the lunar surface brightened by reflected light and visible from Earth ranges through a distinctive cycle, as illustrated in the numbered phases in the series of photographs below right. It begins with the dark new moon *(1, diagram above)*, progresses through a waxing crescent *(2)*, the half-lighted first quarter *(3)*—at which point the Moon has completed one-fourth of its circuit—a waxing gibbous *(4)*, and the full moon *(5)*. The satellite then appears to dwindle during the waning gibbous phase *(6)*, the half-moon at third quarter *(7)*, and the waning crescent *(8)*, before once more going dark to start the cycle anew.

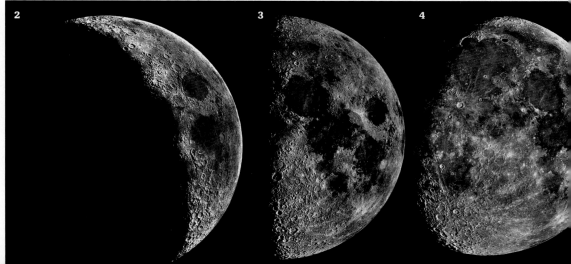

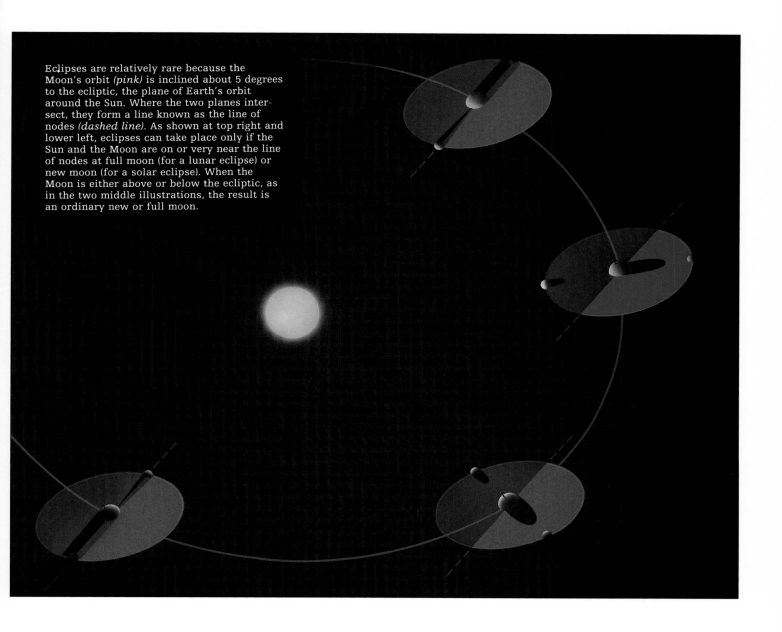

Eclipses are relatively rare because the Moon's orbit *(pink)* is inclined about 5 degrees to the ecliptic, the plane of Earth's orbit around the Sun. Where the two planes intersect, they form a line known as the line of nodes *(dashed line)*. As shown at top right and lower left, eclipses can take place only if the Sun and the Moon are on or very near the line of nodes at full moon (for a lunar eclipse) or new moon (for a solar eclipse). When the Moon is either above or below the ecliptic, as in the two middle illustrations, the result is an ordinary new or full moon.

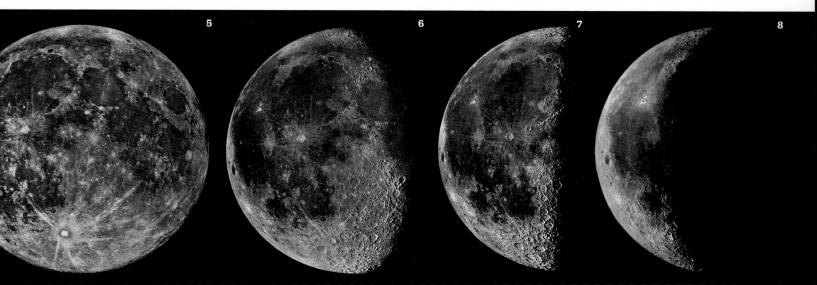

5   6   7   8

it is the interposition of the earth that makes the eclipse, the form of this line (the earth's shadow upon the moon) will be caused by the form of the earth's surface, which is therefore spherical."

Although Arab scholars and observers built upon the foundation laid by the Greeks, Christian authorities largely neglected it for centuries after the classical age waned. Medieval Christians, whose faith demanded a heaven without blemishes, tried to account for the Moon's visibly splotchy face by maintaining that God had hung the Moon like a polished mirror in the sky to reflect the image of the Earth's surface. But the invention of the telescope at the beginning of the seventeenth century doomed any further attempts to portray the Moon as a pristine sphere—and transferred the debate about its nature from the realm of faith to the halls of science.

A PICTURE OF IMPERFECTION

When Galileo refined the telescope as an astronomical instrument at his workshop in Padua in 1609 and trained his device on the Moon, he beheld a decidedly flawed body, strewn with prominences and depressions. He noted cuplike scars, reminiscent of the Grecian wine vessels called kraters, and discerned as well the pale highlands and the dark, irregularly shaped lowlands that looked to him like maria, or seas. Not content with one or two sightings, he returned to the stark lunar landscape night after night to catalog its features, which he recorded in words and in evocative images. The bold pen-and-ink sketches and subtle sepia washes that grace the pages of his notebooks constitute the first realistic lunar portraits. On completing his survey, he echoed the opinion that had caused Anaxagoras to be exiled from Athens 2,000 years earlier: The Moon resembled the home planet. In *Siderius Nuncius,* or *The Starry Messenger,* published in 1610, Galileo wrote, "I feel sure that the surface of the Moon is not perfectly smooth, free from inequalities and exactly spherical, as a large school of philosophers considers, but that, on the contrary, it is full of inequalities, uneven, full of hollows and protuberances, just like the surface of the Earth itself."

This suggestion met with protests from traditionalists, including the Florentine philosopher Ludovico delle Colombe, who strained to preserve the Moon's spotless reputation by insisting that its irregularities were buried beneath a transparent, crystalline sheath whose outer surface was perfectly smooth. This argument did little to undermine Galileo's case, but his pioneering lunar survey was soon subject to needed revisions by astronomers who improved on his efforts. Galileo's illustrations were not altogether accurate; he mislocated and distorted craters, exaggerating here and there out of carelessness or for effect. His compeer Johannes Kepler, who read Galileo's *Starry Messenger* and borrowed from it for his own opus on the Moon in 1634, succeeded in distinguishing between two types of craters, one in which a characteristic circular embankment was punctuated by a central peak, and another in which the embankment ringed a broad, shallow basin. However, in attempting to explain such lunar formations, Kepler proved less astute. The Moon's impressive craters and ramparts, he ventured, had been sculpted by its inhabitants in massive construction efforts, like the pyramids of Egypt or the Great Wall of China.

Together, the observations of Galileo and Kepler paved the way for the systematic study of the Moon's surface, called selenography after the Greek moon goddess Selene. Lunar maps proliferated through the seventeenth century, resulting in some confusion as each succeeding cartographer ignored existing names for lunar landmarks in favor of his own pet titles. Belgian astronomer Michel Florent Van Langren (who, following the custom of the age, went by a Latinized version of his surname, Langrenus), initiated the process in 1645, when he produced a map on which prominent features were chris-

CLOSEUPS OF A FAMILIAR FACE

Along with all of his other achievements, Galileo Galilei became the first lunar cartographer to show the moon in any detail when he made a series of sketches of Earth's satellite in 1609. Today, powerful telescopes, robotic space probes, and human visitors have succeeded in capturing much of the lunar surface in great photographic detail. But as the illustrations on the following pages show, the earliest selenographers, as those who study the Moon's surface are called, were also determined to produce comprehensive and accurate visual records—although imagination sometimes filled in what instruments of the day could not.

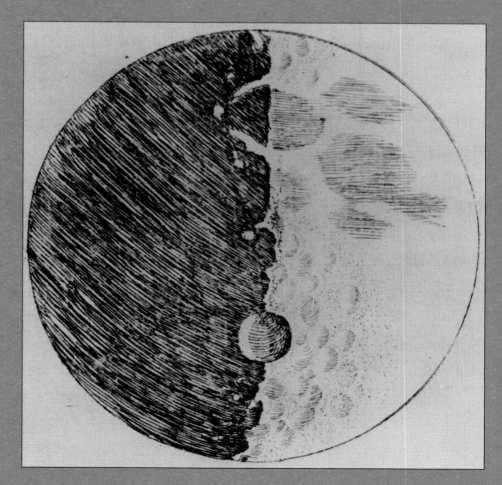

1609 Based on one of Galileo's lunar drawings, this engraving shows a ring of mountains around a feature now known as the Mare Imbrium *(top)* and what appears to be an exaggerated depiction of the crater Albategnius *(bottom).*

tened after crowned heads and famous men—himself included. Just two years later, the gifted Johannes Hewelcke, or Hevelius, who surveyed the Moon with unprecedented precision from his personal observatory in Danzig (now Gdańsk), published a lunar atlas that assigned to most landmarks the titles of geological features on Earth, although Langrenus's own namesake survived Hevelius's revisions.

A few years after that, Hevelius's nomenclature was in turn swept away by Giambattista Riccioli, a Jesuit priest from Bologna whose lunar cartography left something to be desired but whose placenames proved memorable and endured. He chose evocative labels for the various lowlands, including the Mare Tranquillitatis, or Sea of Tranquillity, and the vast Oceanus Procellarum, or Ocean of Storms. Riccioli's fanciful titles took hold among sele-

1651 Giambattista Riccioli introduced the system of lunar nomenclature that is still in use today, with the publication of a Moon map he devised with fellow Jesuit Francesco Grimaldi. In the Oceanus Procellarum, or Ocean of Storms, he named bright craters for Copernicus, Kepler, and Galileo, and two nearby craters *(above, far left)* for Grimaldi and himself.

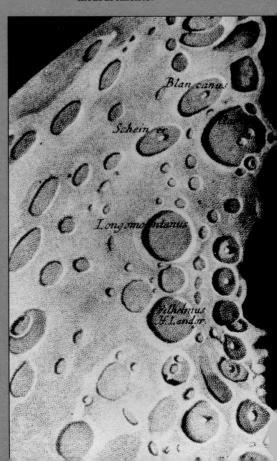

1750 Due to the engraver's error, this mezzotint by German astronomer Johann Tobias Mayer of a region in the lunar highlands is a reverse view—the craters should be on the right of the terminator line between the dark and light sides. Mayer made significant technical contributions to selenography by incorporating a micrometer, mounted inside his telescope, for precise measurements.

nographers of various nations, who thus acquired an invaluable tool—a common set of terms.

The next critical advance in lunar mapping came in the middle of the eighteenth century, when Johann Tobias Mayer, a cartographer in Nuremberg, overlaid his charts of the Moon with a network of coordinates. In this and in the sheer accuracy of his observations, Mayer carried selenography to new heights. He produced two full maps of the Moon's face, seven and a half inches and eighteen inches in diameter, in which the maria figured prominently and thick-rimmed craters were crisply evident, as were the bright rays emanating from many craters and the sinuous lines that snaked across others. Mayer gauged these fine details with the aid of a micrometer, a cross-hair measuring device mounted inside the telescope and controlled by a calibrated

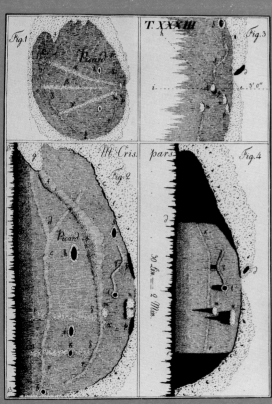

1788 Johann Hieronymus Schröter made a series of increasingly detailed drawings of the Mare Crisium, zeroing in on cracklike features known as clefts or rilles. Schröter produced hundreds of such drawings, studying the Moon's features at varying angles of illumination to derive precise measurements of crater diameters and altitudes.

1837 Wilhelm Beer and Johann Heinrich von Mädler produced the first large-scale lunar map, a section of which is shown here, along with a text describing each feature extensively. So definitive was the map that it served as a standard for the rest of the century.

screw. Equipped with the same tool, Mayer's able compatriot Johann Hieronymus Schröter later scrutinized craters even more closely and determined that a one-to-one relationship exists between the volume of the material mounded in a crater's outer walls and that of the depression encircled by them, a maxim known as Schröter's Rule. In a more speculative vein, he noted a resemblance between lunar craters and volcanic formations on Earth and suggested that the Moon's pocked face was the result of volcanic activity. He thus injected himself into a debate that had been brewing for more than a hundred years and that would persist into the twentieth century, as astronomers and geologists attempted to identify the powerful force that had ravaged the Moon's surface.

MIMICKING THE MOONSCAPE

One of the first to ponder the origins of lunar craters was Robert Hooke, a distinguished inventor and mathematician and a leading member of the Royal Society of London for Improving Natural Knowledge. In the mid-1600s, Hooke became fascinated by the Moon's surface irregularities and their possible cause. A firm believer in the experimental method, he tried to recreate the process in the laboratory, working with a thick, wet mass of plaster intended to simulate the surface of the Moon at an early, impressionable stage. When a pot of boiling plaster was removed from the fire, he reported, "the whole surface, especially that where some of the last Bubbles have risen, will appear all over covered with small pits, exactly shap'd like these of the Moon, and by holding a lighted Candle in a large dark Room, in divers positions to this surface, you may exactly represent all the *Phaenomena* of these pits in the Moon, according as they are more or less inlightned by the Sun." Hooke also threw musket balls and pellets of mud into another viscous concoction and reported that the missiles left craterlike marks. However, he found it difficult to imagine how any such projectiles could have struck the Moon, since he shared the prevailing opinion that interplanetary space was void.

Hooke's dabbling convinced him, and many other scientists, that lunar craters were volcanic in origin, a conclusion reinforced by their apparent resemblance to the vents of terrestrial volcanoes, encircled by rims of solidified magma. The idea of lunar volcanism proved so compelling that seasoned observers began to infer the existence of active vents on the Moon from hazy impressions. No less a figure than William Herschel—the brilliant German-born astronomer who discovered Uranus from his observatory in Bath, England, in 1781 and went on to spot two moons around that planet—reported sighting volcano-like eruptions on the Moon.

Despite Herschel's high standing among his fellow astronomers, some wondered whether he had correctly interpreted what he perceived, and whether in fact volcanoes were responsible for lunar cratering. A series of well-documented meteorite falls in Italy, England, and France in the late eighteenth and early nineteenth centuries raised doubts about the proposition that interplanetary space was empty. As yet, no crater on Earth had been linked to an object plummeting from the skies, but by the 1820s at least one authority, the German astronomer Franz von Paula Gruithuisen, was willing to venture publicly that lunar craters were the result of cosmic bombardment. Unfortunately, Gruithuisen also claimed to have observed signs of habitation on the Moon—including the apparent outlines of cities—and his impact ideas received little serious attention as a result. Indeed, some reputable scientists insisted that the recent meteorite falls, far from buttressing the impact origin of lunar craters, offered proof to the contrary. The meteorites, they explained, were volcanic ejecta from new craters forming on the Moon. This ingenious argument was later discounted when astronomers learned more about the chemistry and orbits of meteorites and traced them to distant regions of the Solar System. But the role of past volcanism in sculpting the lunar surface became scientific gospel, thanks in large part to German astronomer Johann Heinrich von Mädler and his patron and colleague, Wilhelm Beer, who endorsed the argument in their definitive study of the Moon, published in 1837 along with the most accurate lunar map to date. Earth's satellite, they concluded, "is to all intents an airless, waterless, lifeless, unchangeable desert, with its surface broken by vast extinct volcanoes."

UNEARTHING FRESH EVIDENCE

Notwithstanding such pronouncements, those scientists who compared lunar craters systematically with their terrestrial counterparts found it hard to believe that they shared a common origin. Among the skeptics was Grove Karl Gilbert, an expert American geologist who spent years scouring the rugged terrain of the western United States, encountering more than a few volcanic craters along the way. A man of wide scientific interests, Gilbert was familiar with the volcanic theory of lunar-crater formation. In December 1892, a few months after his retirement as chief of the U.S. Geological Survey, Gilbert drew on his surveying experience and other research to dissect that hypothesis in an address to the Philosophical Society of Washington.

Gilbert brought an array of evidence to bear against the volcanic theory. For one thing, he pointed out, lunar craters greatly outnumbered terrestrial ones, suggesting that the Moon must have been far more active than Earth to spawn so many volcanic monuments; yet by all other appearances, the Moon had long been geologically dead, undergoing none of the upheaval that had lent such variety to the surface of Earth. For another thing, the ten largest lunar craters were, on average, twenty-five times wider than the ten largest volcanic vents on Earth, a discrepancy that was roughly four times what one might expect, even after accounting for the lower gravity on the Moon (about one-sixth that on Earth). More important, lunar craters were deep depressions, while the typical volcanic craters on Earth—the so-called Vesuvian type—were bowls that descended steeply from the volcano's rim but bottomed out above the level of the surrounding countryside. "Ninety-nine times in one hundred," Gilbert emphasized, "the bottom of the lunar crater lies lower than the outer plain; ninety-nine times in a hundred the bottom of the Vesuvian crater lies higher than the outer plain." Lunar craters, he concluded, were caused not by the upwelling of volcanic material but by gouging impacts.

To bolster his contention that the Moon's cavities were produced by collisions with swiftly moving objects from space, Gilbert detailed the results of cratering experiments he had recently conducted. While in New York City as a guest lecturer at Columbia University in 1891, he had spent many evenings in his hotel room miming his eminent British forerunner, Robert Hooke. In place of the plaster used by Hooke, Gilbert relied on mud slurries to simulate the lunar surface and tossed balls of clay into the mixture to replicate the action of meteorites. The impressions that resulted left him with little doubt as to the validity of the impact theory. As he insisted in his address: "If any projectile be made to strike any plastic body with suitable velocity, the scar produced by the impact has the form of a crater." He hastened to add that the Moon could be plastic without being soft or malleable: The heat produced by a large stony fragment striking its surface would be enough "not only to melt the fragment itself, but also to liquefy a considerable tract of the rock mass by which its motion was arrested." Particularly fierce collisions would send molten rock splattering out in all directions from the point of impact, thus accounting for the rays emanating from some craters. (Lunar reconnaissance would later confirm that the rays contained once-molten material, along with dust and other powdery ejecta.)

The theory that the Moon's craters are volcanic in origin failed to account for an underlying difference between lunar craters and those formed by volcanism on Earth, as illustrated in the diagrams accompanying the photographs below. Terrestrial volcanic calderas such as Mexico's Cerro Colorado *(below, left)* typically have central lava plains that lie above the surrounding terrain. By contrast, the floors of nearly all lunar craters *(right)* are sunken, indicating that they were excavated by impacts.

As cogent as Gilbert's arguments were, they made little impression at the time. He had some difficulty explaining why the Moon's craters were predominantly circular. His low-velocity experiments with mud had indicated that an object coming in at an oblique angle would yield an oval scar, and the scarcity of oval craters on the Moon led him to propose a complicated set of circumstances whereby the majority of impacting objects plummeted directly down on the lunar surface rather than arriving at various inclinations as chance would dictate. Also, few in Gilbert's audience were prepared to accept the idea that space was ever as riddled with dangerous debris as the impact theory implied. Lacking any proof of such devastating bombardments on Earth, astronomers saw no reason to accept them on the Moon.

As it happened, Gilbert had come close to discovering such proof not long before he delivered his address. Early in 1892, he had surveyed a gaping crater in northern Arizona that resembled those on the Moon. More than 4,000 feet wide and 600 feet deep, the pit was ringed by iron fragments—an indication that an iron-rich meteorite had carved the depression. Gilbert had hoped to confirm that explanation by detecting signs of a rich lode of iron beneath the floor of the crater. But his instruments revealed no such deposit, and he concluded reluctantly that the pit had been formed by an eruption of underground steam. However, later excavations at the site turned up compelling evidence of a meteorite fall there—including large deposits of powdered silica, indicating that sandstone had been pulverized by a tremendous blow. By the early 1920s, earth scientists were coming to the conclusion that the crater was forged by impact.

The Arizona crater struck at least one interested authority as confirmation of the impact origin of lunar craters. In 1921, the German geophysicist Alfred Wegener, architect of the theory of continental drift, published a paper on lunar craters that reformulated the impact hypothesis in light of the latest evidence and did much to resuscitate Gilbert's arguments. Wegener reported the results of his own cratering experiments using cement—tests that bore out Gilbert's conclusions. Wegener's experiments and his review of lunar cartography convinced him that the elevated rims of the Moon's craters were simply mounds of debris displaced from the gouged-out basins by the force of impact. "A lunar crater is not a mountain at all," he emphasized, "because the walls that are thrown up are just enough to compensate for the depression of the floor. If we would pour the material of the walls into the crater we would just fill it." This was Schröter's Rule, but Schröter had been unable to gauge its significance without the geologic evidence that later emerged, showing the maxim to be at odds with volcanic processes and consistent with impact. Wegener, profiting by the latest findings, argued that the only large formations on Earth that conformed to the lunar pattern were the meteorite crater in Arizona and other sizable pits around the globe that appeared to be of similar origin. He concluded that "magnificent meteorite impacts" had occurred repeatedly on Earth, although only the more recent craters had survived; the rest had been obliterated by erosion.

Like Gilbert, Wegener failed to arrive at a satisfactory explanation for the circular configuration of most lunar craters. That major obstacle to acceptance of the impact theory was removed by several researchers who hit independently on the expedient of examining bomb craters to see what light they shed on the problem. The first such studies were done during the First World War, before Wegener published his paper, but conclusive evidence was not amassed until the Second World War. The research indicated that an exploding missile yields a roughly circular crater even when it strikes the ground at a fairly low angle. Although both Gilbert and Wegener had recognized the explosive nature of lunar impacts, they had not reproduced that effect in their low-velocity experiments, and the results had led them astray.

Although the bomb-crater evidence strongly supported the impact theory, it took some time for the significance of the disparate findings to register. It fell to American astronomer Ralph Baldwin to sum up the case for the scientific community in 1949 in his lucid and influential study, *The Face of the Moon.* Baldwin's exhaustive comparison of lunar depressions with natural and artificial craters on Earth argued convincingly for impact. "The only type of crater that corresponds to the ones on the moon is the simple explosion pit formed by a single application of explosive power," he concluded. "Such a pit always has the same general form, whether it is produced by a bomb, a shell, a military mine or a meteorite."

Baldwin's verdict was largely accepted, although pockets of resistance to the impact hypothesis persisted in scientific circles for nearly two decades. By the time American astronauts prepared to journey to the Moon in 1969, virtually all astronomers agreed that its craters were the product of bombardment. One vexing lunar riddle had at last been answered. But another question remained to be settled, a problem that went beyond the Moon's surface blemishes to the very core of its identity. How had this solitary satellite formed? Like the puzzle of the craters, the deeper question had elicited conflicting theories over the years. And the long debate had brought the experts no closer to a solution. Not until astronauts walked on the Sea of Tranquillity and returned to Earth with revealing samples of the lunar crust would scientists begin to plumb the mystery of the Moon's birth.

THE HUNT FOR BEGINNINGS

Astronomers and geologists began to debate the origins of the Moon in the late 1800s, even as the wrangle over cratering was heating up. By that time, observers had identified several curious attributes of the Moon that seemed to bear directly on its origins. One was the Moon's ample size relative to Earth: The satellite was nearly one-fourth the diameter of the planet. Although nineteenth-century astronomers could only venture rough estimates of the size of distant moons by gauging their brightness, there appeared to be no other satellites that loomed so large in comparison to their planets. Somehow, Earth—a fairly small sphere on the planetary scale—had spawned a prodigious offshoot. On the other hand, the Moon was not nearly as dense as Earth.

Two bodies hovering as close to each other in space as Earth and the Moon would presumably have formed under similar conditions, and should thus be similar in makeup. Yet calculations of the Moon's density, derived by comparing its volume to its mass as revealed by various orbital characteristics, showed the satellite to be made of lighter stuff. Its density amounted to a mere 60 percent of Earth's. Some scientists concluded that the Moon's interior harbored a smaller quotient of heavy minerals such as iron and nickel, but they could not explain why.

A third factor any lunar origin theory would have to account for was the angular momentum of the Earth-Moon system—a complicated formula based on the mass of the two bodies, the distance between them, and the rate at which Earth rotates on its axis and the Moon rotates about Earth. According to a principle laid down by Isaac Newton, the momentum of a given system must remain constant in the absence of an external force. If, for example, the rotation rate of the Earth decreases, then the distance of the Moon from Earth must increase at the same time to conserve angular momentum. (A simple illustration of this principle is offered by the twirling figure skater whose spin rate decreases as she extends her arms.) As it happens, the tidal bulges produced on Earth by the Moon's gravity *(page 26)* are in fact slowing the planet's spin rate by a fraction of a second each year, and the Moon is compensating by drifting away ever so slowly. The scientists who first pondered the origins of the Moon knew by comparing eclipses and other celestial events at long intervals that Earth was spinning more slowly than in the past, and they tried to fit this curious piece of information into the puzzle.

Appropriately, the first detailed theory of the Moon's evolution was the work of George Howard Darwin, son of the great naturalist who probed the origin of species. Like his father, Charles, George Darwin indulged other interests for a time before settling into a scientific career. After graduating with honors from Trinity College, Cambridge, in 1868, he studied law for six years before returning to Trinity to teach astronomy and pursue his specialty: the mathematical analysis of orbiting bodies. Pondering the fact that Earth's rotation rate was slowing, Darwin correctly inferred that the Moon must be receding from Earth at the same time to conserve angular momentum. He next surmised that just as it was possible to predict from this tendency what the Moon's position and Earth's period would be in the future, it was equally possible to project the changes in lunar orbit and terrestrial rotation back in time. Carrying out the laborious calculations, he reached a point more than 50 million years ago when Earth spun on its axis once every five hours and the Moon hovered just 6,000 miles above the planet's surface. Extending the process to an even earlier stage, Darwin concluded that the Moon must have been an offshoot of a viscous young Earth that was spinning so rapidly that centrifugal force had caused it to bulge out markedly at the equator. The Sun's gravity then tore away a large glob of the protruding material, and the glob ultimately solidified to become the Moon.

This scenario, variously known as the daughter, escape, or fission hypothesis, sounded eminently sensible to many scientists when Darwin first proposed it in 1878. One of the theory's strong points was its neat explication of the Moon's low density. The young planet's lighter materials would have migrated to the bulges during its hectic pirouette, so that the glob cast off would be considerably less dense on average than the matter left behind. More than one advocate of the theory went so far as to identify the Pacific Ocean basin as the site of the ancient separation.

Despite its initial appeal, the fission theory eventually lost luster. One of the first objections raised to Darwin's hypothesis involved a principle articulated in 1848 by French mathematician Édouard Albert Roche, who had calculated that within a given distance of a planet, its gravitational influence would keep dispersed material from clumping together to form a moon. Plainly, any viscous glob that spun off the Earth would be inside this so-called Roche limit; prevented from accreting, the jettisoned material would orbit the planet as a ring of particles. Darwin conceded the point, but insisted that even such a ring would produce sufficient tidal bulging on Earth to help slow down the planet; this in turn would induce the orbiting material to drift away until it was outside the Roche limit and could consolidate into a moon.

Other arguments against fission proved more difficult to refute. In order for the planet to have bulged out through centrifugal force, later calculations revealed, its angular momentum would have had to be four times greater than that of the Earth-Moon system today, a baffling discrepancy in light of Newton's conservation principle. And even at that improbably high level of momentum, the young Earth would have had to be composed entirely of molten material no more viscous than water to be subject to fission, a possibility that struck geologists as remote given the present makeup of Earth. Moreover, radiometric dating techniques developed in the early twentieth century indicated that Earth was far older than Darwin or any of his contemporaries had imagined. Eventually, Earth's age was fixed at 4.6 billion years, while revised estimates for the time

In 1879, George Howard Darwin *(left)*, son of Charles Darwin and an esteemed scientist and mathematician in his own right, introduced his century's most widely accepted theory of the Moon's genesis. According to the so-called fission hypothesis *(below)*, the Sun's gravity compounded the bulging of the fluid primordial Earth as it rotated rapidly on its axis, so that a portion of the planet was flung off into space to form the Moon.

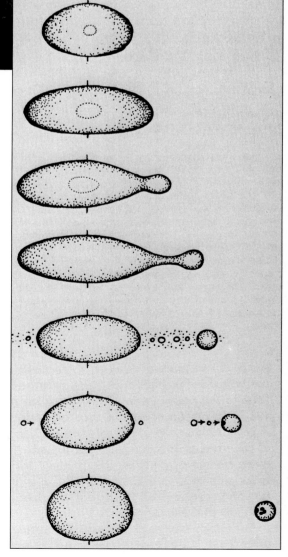

scale of the fission process traced the Moon's separation from the planet back 2 billion years at most—by which time Earth would presumably have firmed up, rendering the scenario all the more improbable.

The evident shortcomings of the fission hypothesis fueled scientific interest in competing lunar origin theories, including one that had been advanced in general terms a few years before George Darwin worked out his detailed model. This alternate hypothesis was known as the coaccretion, or "sister," theory, since it portrayed the Moon more as Earth's younger sibling than as its offspring. The theory, which held that the Moon condensed outside the Roche limit from a disk of gas and dust encircling the proto-Earth, was a logical extension of the argument that the planets of the Solar System had formed from a gaseous nebula swirling about the Sun. Elaborations of that nebular hypothesis described how, over time, innumerable collisions, first between tiny particles of dust and gas, and later between pebble- to boulder-size bodies, created chunks of matter called planetesimals, which eventually accreted to form the planets. In much the same way, suggested Édouard Roche in 1873, the Moon might have accumulated from a contingent of planetesimals bound by Earth's gravity. Roche advanced this theory only tentatively, and with good reason, for it did little to account for the Moon's peculiar features. For one thing, it failed to explain why Earth, alone among the small, rocky inner planets, should thus have acquired a large satellite. Only the outer gas giants Jupiter, Saturn, and Uranus presented a strong case for this type of moon formation *(page 33)*.

A third origin theory dispensed with the troublesome proposition that the Moon had developed near Earth. Called the capture hypothesis, it was proposed in 1909 by American astronomer Thomas Jefferson Jackson See. See contended that the Moon had taken shape not in the inner Solar System but farther out, beyond the orbit of Uranus. There, billions of miles from the Sun, whose heat stripped the developing inner planets of their lighter elements, conditions favored the formation of a body significantly less dense than Earth. When this new object first accreted, it orbited the Sun. But as it plowed through gas, dust, and other cosmic debris in its path, it lost energy and its orbit shrank. Eventually, it passed close enough to Earth to fall under the gravitational sway of the planet.

That planets could in fact capture satellites had been suggested by recent discoveries of tiny satellites circling Jupiter in highly elliptical orbits, the likely path for a small object that strayed obliquely into the gravitational clutches of a giant planet *(pages 34-35)*. But the Moon presented a different picture. It was quite large relative to Earth, and its orbit was nearly circular, raising serious problems for See's hypothesis. No less difficult to explain through capture was why a Moon that had been drawn in toward Earth should now be edging away. See, who was something of an eccentric, simply insisted that the pundits had misinterpreted the evidence, and that the Moon was actually inching nearer to Earth—an unfounded assertion.

Within a few decades, however, the capture hypothesis was revived by

THE DISTORTING EFFECTS OF GRAVITY

The ocean tides that swell and recede daily on Earth are just one manifestation of the gravitational relationship between orbiting bodies. A planet and its satellites—or the Sun and its planets—exert differential gravitational forces on each other. That is, because gravity decreases with distance, the forces are stronger on the facing hemispheres, weaker on the bodies' far sides *(below)*, distorting the two bodies' shapes and modifying their orbital characteristics.

Because fluids are more malleable than solids, ocean tides are the most visible effect of tidal tugging *(bottom and right)*. But the solid parts of orbiting bodies also respond, bulging into somewhat football-like shapes. In the Earth-Moon system, the Moon lags behind Earth's tidal bulges as the two bodies orbit. This lag, combined with the differential forces, generates internal tidal friction that has slowed the rotational spin of both bodies as energy dissipates in the form of heat. Every century, Earth's axial rotation period slows two-thousands of a second, lengthening the day by that amount. The Moon, subject to twenty times the tidal force felt by its planet, has slowed until it spins at exactly the same rate that it orbits Earth. As a result of this synchronous rotation, it shows the same face to Earth at all times—a fate shared by the majority of satellites in the Solar System.

Tidal forces are differential gravitational forces experienced at different points on bodies in orbit around one another. All points on both worlds are subject to gravity, but the intensity of the attraction varies with distance from the other body. The leading edges feel the strongest pulls *(represented by long arrows)*, the back sides suffer the least *(short arrows)*, and the center, top, and bottom receive a moderate tug.

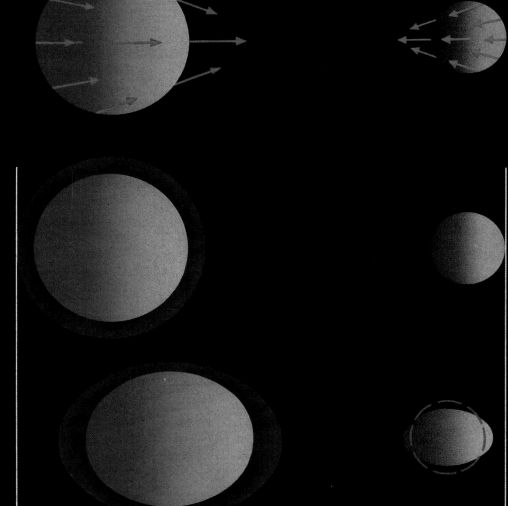

If Earth and its satellite experienced no tidal distortion, they would be perfectly round and keep a slightly greater distance from each other than they actually do *(right)*. Because of tidal interaction, however, the bodies are pulled toward each other *(bottom)* and—because the facing surfaces are pulled more vigorously than the rear—their spheroid shapes are distorted, as is Earth's water. On average, water on the side near the Moon is pulled more than Earth's center, causing a high tide. And water on the far side is pulled less than Earth's center, resulting in another high tide there.

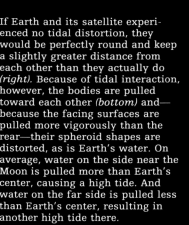

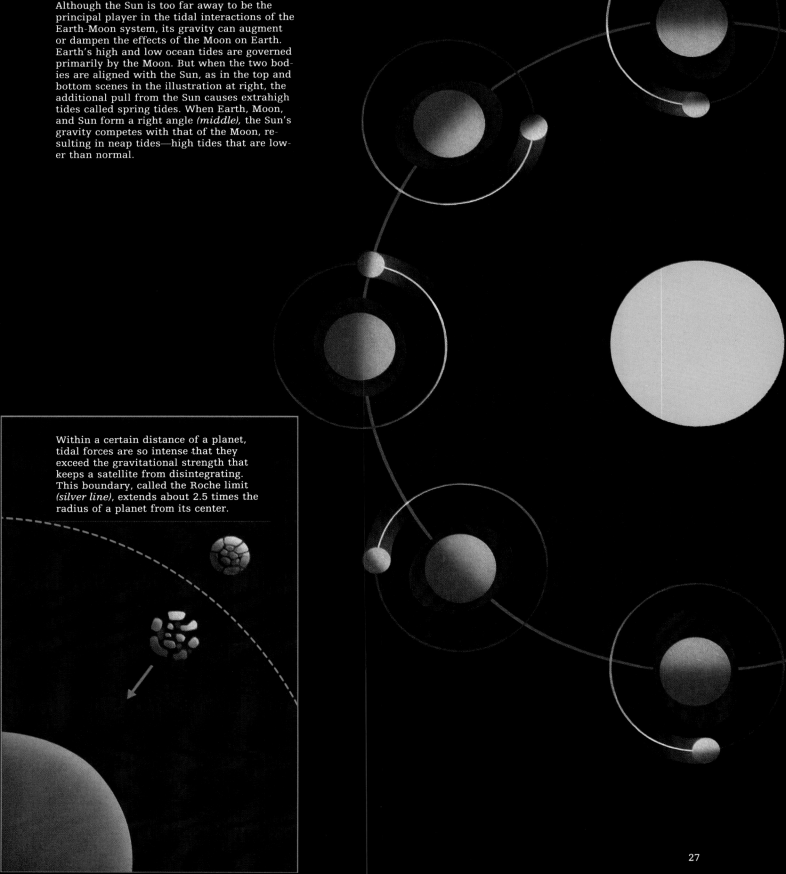

Although the Sun is too far away to be the principal player in the tidal interactions of the Earth-Moon system, its gravity can augment or dampen the effects of the Moon on Earth. Earth's high and low ocean tides are governed primarily by the Moon. But when the two bodies are aligned with the Sun, as in the top and bottom scenes in the illustration at right, the additional pull from the Sun causes extrahigh tides called spring tides. When Earth, Moon, and Sun form a right angle *(middle),* the Sun's gravity competes with that of the Moon, resulting in neap tides—high tides that are lower than normal.

Within a certain distance of a planet, tidal forces are so intense that they exceed the gravitational strength that keeps a satellite from disintegrating. This boundary, called the Roche limit *(silver line),* extends about 2.5 times the radius of a planet from its center.

An Ascending Scale of Impacts

Planetary scientists have made sense of the bewildering profusion and variety of lunar craters by classifying them according to their size and shape, with distinct features corresponding to different magnitudes of meteoritic impact. So-called simple craters *(near right)* are neat bowl-like cavities with circular rims whose diameters are generally less than nine miles. However, larger craters, forged by explosions of greater power, tend to develop more elaborate structures. Their collapsed walls are terraced rather than smooth, and their floors, strewn with debris, are level rather than curved. Such larger, rougher craters can all be classified as complex, but three ranks have been identified within that broad category.

At widths starting around twelve miles, lunar craters are marked by central peaks *(second from left)*, formed by the uplifting of bedrock brought on by the drop in pressure as the impacting meteorite vaporizes. In craters ninety miles or more wide, the uplifting is broader and more diffuse, producing an inner cluster, or ring, of mountains *(second from right)*. The grandest impact structures of all, known as multiring basins, extend for several hundred miles and are marked by concentric rings of steep scarps and mountains *(far right)*.

scientists who considered the fission and coaccretion scenarios even less plausible and who argued that over vast stretches of time, tidal effects could alter the paths of captured satellites dramatically, explaining the Moon's present orbital characteristics. Advocates of the competing models made similar attempts to adapt their theories to the facts, and their efforts succeeded to the extent that none of the three main alternatives could be dismissed out of hand. By 1961, when President John F. Kennedy proclaimed America's intention to send astronauts to the Moon, the mystery of that body's origins appeared as tangled as ever. Scientists could only hope that the planned Apollo moonwalks—along with ongoing American and Soviet efforts to probe the satellite with unpiloted spacecraft—would provide the clues needed to solve the stubborn problem.

The urge to explain the Moon's birth was just one of the scientific motives behind the Apollo flights. Evidence was accumulating that after a volatile early phase marked by frequent impacts, the surface of the Moon had remained essentially unchanged while Earth underwent billions of years of geologic evolution. If this view was accurate, then the Apollo program would represent more than a bold voyage across space. It would constitute a journey back in time, to a revealing relic of the formative era of the Solar System.

GUESSING THE MOON'S AGE

The scientist who did more than any other to promote the Apollo program as a field trip into the cosmic past was University of Chicago professor Harold C. Urey, a celebrated chemist who came late to the field of astronomy and proceeded to challenge the assumptions of some of its leading lights. In 1934, at the age of forty-one, Urey had shared the Nobel prize in chemistry for his

Theophilus. This 63-mile-wide crater near the Mare Nectaris has terraces and central peaks, typical of a modest-size complex crater.

Schrödinger. Roughly three times the width of Theophilus, this crater in the south polar region has a loosely defined inner ring of peaks.

Mare Orientale. This 560-mile-wide crater, a classic multiring basin, has been pocked by smaller craters since its formation.

discovery, with two colleagues, of heavy hydrogen, or deuterium. During World War II, he headed the Columbia University team that helped to develop the atomic bomb under the auspices of the secret Manhattan Project. Dismayed by the devastating results of that effort, Urey became a vocal pacifist after the war and abandoned nuclear research for other pursuits, including an ambitious inquiry into the origin of the Solar System.

Early on in that quest, Urey came to regard the Moon as a case study of conditions in the early Solar System. That the Moon had undergone little evolution since its formation appeared evident in the crisp outlines and coherent profiles of many of its craters and ridges. The young Earth must have presented a similar picture, Urey reasoned, but it had been altered dramatically not only by wind and rain but by internal processes such as volcanic eruptions and earthquakes, caused by heat generated in the planet's core. The Moon, by contrast, appeared to have maintained its original identity as a geologically inert body with too little internal heat to alter the surface.

These observations dovetailed with Urey's central thesis—that the planets and their satellites had accreted from a relatively cold nebular cloud of gas and dust rather than a hot one, as some cosmologists argued. Evidence of this lay in the Moon's perceptible tidal bulge. If the Moon had once been hotter and more plastic, Urey contended, its interior would not have been stiff enough to support the bulge, which would have flattened out over time as the Moon drifted farther from Earth and its tidal pull. This cold-Moon theory left Urey at some pains to explain the dark surface of the lunar maria, which most scientists attributed to basaltic lava flows. Perhaps the molten material was simply the result of intense surface heating caused by the tremendous impacts that gouged out the basins, he suggested, or perhaps the maria were not lava beds at all but just what their name implied—ancient seabeds, encrusted with the residue of primitive organisms. If that was the case, Urey wrote, "the Apollo program should bring back fascinating samples which will teach us much in regard to the early history of the solar system, and in particular with regard to the origin of life."

In this, as in his suggestion that the Moon's interior might never have been molten, Urey turned out to be mistaken. But he was correct in viewing the Moon as a cosmic relic. That fact was confirmed in the early 1960s by a team of geologists led by Eugene Shoemaker, an Arizona geologist with the U.S. Geological Survey. Relying on precise lunar images and direct telescopic observation, Shoemaker and his colleagues applied the technique of stratigraphy— the analysis of Earth's geologic strata—to sort out the layers of material on the Moon's surface and construct a time scale for the events that deposited them. A crater that overlay and partly obliterated another crater, for example, had obviously been carved at a later date.

Just under a mile wide and a thousand feet deep, Hadley Rille snakes through the eastern part of the Mare Imbrium. Scientists think that molten lava once coursed through this chasm, which may have been sculpted by the flow or by an earlier surface fracturing.

Employing this system, Shoemaker divided the features on the lunar surface into five groups, or stages, ranked from oldest to youngest and named for prominent landmarks exemplifying the various stages: pre-Nectarian, Nectarian, Imbrian, Eratosthenian, and Copernican. The oldest areas were well worn, having been eroded over the eons by sporadic major bombardments and the constant flow of micrometeorites—specks of cosmic debris hurtling through space. The youngest impressions were fresh and crisp, such as the sharply defined crater Copernicus in the Ocean of Storms. Most regions of the Moon were layer cakes of succeeding stages. For example, the Mare Imbrium, or Sea of Showers, in the Moon's northwest quadrant, contained many small and relatively recent craters superimposed on a dark surface of intermediate age, which in turn covered a broad basin gouged out of the Moon's crust by an ancient massive impact.

As precise as this scheme was, it could offer only relative dates for lunar events. However, Shoemaker suggested a technique for determining the absolute age of the various features. Geologists could estimate cratering rates on Earth, he noted, by ranking craters according to size and dating them to determine how many meteorite falls of that magnitude had occurred within a given area over time. An event such as the one that carved the big Meteor Crater in Arizona, for instance, had taken place on the North American continent roughly once every 10,000 years. Assuming that the Moon was subject to the same rate of bombardment, Shoemaker calculated from the profusion of such scars standing out crisply on the lunar surface that the Copernican

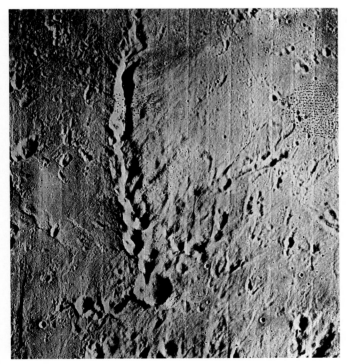

This secondary crater chain was
etched by ejecta spewed out by the
impact that carved a prominent
nearby crater named Copernicus;
large chunks of debris spattered
back to the surface here in a line.

epoch—the most recent cratering period—had in fact lasted approximately 2 billion years. The stage before that, the Eratosthenian, extended back nearly to the period of Earth's formation, 4.6 billion years ago. From this, Shoemaker deduced two things—that the Moon was indeed as old as Earth, and that a great deal of lunar cratering, including the massive impacts that carved the maria, had occurred in a short span after the satellite's formation. "In their earliest stages the moon and the earth were encountering in space debris left over after the consolidation of the planets," Shoemaker explained. "It would take about 100 million years for the earth and the moon to sweep up the planetesimals in their paths around the sun." Following this spate of major impacts, the rate of bombardment declined markedly. Erosion, sedimentation, and mountain building gradually covered up all but the most recent craters on Earth, while the quiescent Moon retained each succeeding impression.

A SUSPENSEFUL TOUCHDOWN

Shoemaker's calendar for lunar cratering did not solve the mystery of the Moon's birth. But it underscored the scientific importance of the forthcoming Apollo flights and raised hopes that astronauts would return from the Moon with crucial evidence. As late as the mid-1960s, however, there was some doubt as to whether a spacecraft could safely touch down on the surface of the Moon. The potential problem was identified by astrophysicist Thomas Gold of Cornell University, who had been pondering the slow lunar weathering processes that produced the Moon's regolith, or unconsolidated surface material. Gold observed that while there were no winds, rivers, or waves to erode mountain chains on the Moon, there were incessant showers of micrometeorites that would pulverize surface rock, turning it to a fine powder. The powder's tiny constituents were subject to further bombardment by streams of subatomic particles flowing perpetually from the Sun and beyond. Gold speculated that these solar and cosmic particles could convey a charge to the lunar dust motes, causing those on the top layer to hover above the rest, buoyed by electrostatic repulsion. The floating dust on steep slopes would then be tugged by lunar gravity and glissade downhill, settling in low areas that might be targets for future lunar landings.

How deep these pools of powder might be, Gold could not say exactly. But concern arose that spacecraft settling on the Moon might be swallowed as if by quicksand. Mission planners could only hope that the lunar regolith—which consists of rock fragments as well as powder—would prove firm enough to support the weight. Worries about possible dust dunes persisted until February 1966, when the unpiloted Soviet probe *Luna 9* made a safe, soft landing in a light coating of dust on the Ocean of Storms.

This feat constituted another first for the Soviet Luna program, which had begun spectacularly in 1959 with three Moon shots, the last of which had looped around Earth's satellite and beamed back unprecedented views of the far side. Although blurry and indistinct, the photographs revealed few maria, indicating that those plains were largely confined to the near side—a fact that continues to puzzle scientists to this day *(pages 48-49)*. Following the soft landing of *Luna 9* in 1966, the Soviets brought a succession of unpiloted craft down on the Moon over the next ten years to analyze the surface. The last probe in the series, *Luna 24,* landed in the Mare Crisium, or Sea of Crises, on August 18, 1976, drilled down more than six feet, and returned to Earth four days later with the core sample. Such robotic feats helped to unlock the Moon's geologic secrets by surveying ground uninvestigated by the more celebrated Apollo missions.

Those flights represented the triumphant fulfillment of an American lunar campaign that had begun ignominiously in 1961 with a string of unpiloted Ranger missions that went awry. The problems were eventually ironed out, and in 1964, *Ranger 7* crash-landed as assigned on the Mare Nubium, or Sea of Clouds, after transmitting the best closeup photographs of the lunar surface to date. As a result, the name of the region was later changed to the Mare Cognitum, or the Known Sea. In May 1966, the Americans drew even with the Soviets when *Surveyor 1* made a soft landing on the Moon, returning over 11,000 splendid images. Two years later, the first Apollo crews made it into Earth orbit and later lunar orbit. Then in early July 1969, the countdown began for the historic mission that would set humans down on the Moon. Planetary scientists anxiously awaited the results of this first extraterrestrial field trip—and experienced more than a little envy for the pioneers selected to carry it out. "I wish I could go rock-hunting with the astronauts this month," the elderly Harold Urey told a reporter on the eve of the flight of *Apollo 11*. "I think I'd go," Urey added, "even if I knew I could never get back."

THE STORY OF THE STONES

On July 20, 1969, astronaut Neil Armstrong took his great leap for mankind, stepping from the lunar module of *Apollo 11* and planting his feet on the gray, dusty surface of the Sea of Tranquillity. Moments later, he fulfilled the dream

Above, the core of a large protoplanet *(dark purple)* is surrounded by gas and dust. In time, gravitational and rotational forces cause the envelope to expand and flatten into a distinctively shaped protoplanetary nebula *(below)*.

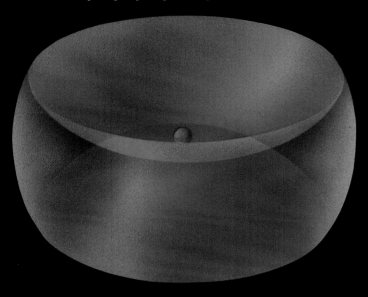

Densest and hottest near the planet, the protoplanetary nebula flares and becomes more diffuse with distance *(above)*. Eventually, solid particles in the protoplanetary nebula—dust and ice—settle toward the midplane, forming a disk *(purple oval, below)* that orbits the planet's equator. Solid bodies begin to build up there and eventually accrete to yield satellites.

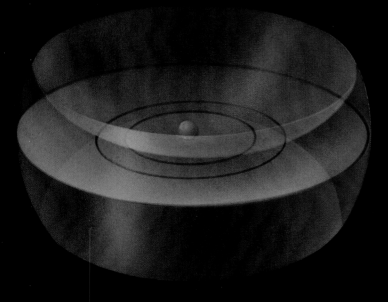

THE VARIED ANCESTRY OF MOONS

For years, scientists pondering the origin of Earth's moon weighed three possibilities: that the satellite formed near Earth through the same gradual process of accretion that produced the planet, that the Moon took shape elsewhere in the Solar System and was captured by Earth when the two bodies crossed paths, or that the Moon came about through fission of the proto-Earth when the rapidly spinning planet threw off a huge glob of viscous material. The flood of data returned by the Apollo missions and other lunar probes eventually led most scientists to dismiss these explanations in favor of a fresh alternative—the giant impact theory, which holds that a collision between the young Earth and a large intruder blasted a vast amount of debris into space to form the Moon *(page 36)*. Yet the earlier theorizing was not in vain. Concepts that failed to account for Earth's satellite have been adapted to explain the origin of moons circling the Sun's other planets.

The accretion process illustrated here, for example, is thought to apply to dozens of moons grouped around the large gas planets Jupiter, Saturn, and possibly Uranus. These tidy satellite families constitute miniature solar systems, with the members moving in nearly circular paths close to their planet's equator. Such moons apparently resulted from an evolutionary sequence much like the one that produced the planets themselves—beginning with a spherical envelope of gas and dust swirling around a center of gravity, leading in time to the formation of an equatorial disk of revolving particles, and culminating with the consolidation in the disk of large orbiting bodies at discrete intervals. To complement such regular offspring, the outer planets have evidently acquired some moons through capture as well, adopting orbiters that strayed into their gravitational fold *(pages 34-35)*.

The mini–solar system is now formed. An intense blast of solar wind—an outburst characteristic of young stars—cleared out the remaining nebula, leaving the moons in neatly spaced circular orbits around the planet's equator.

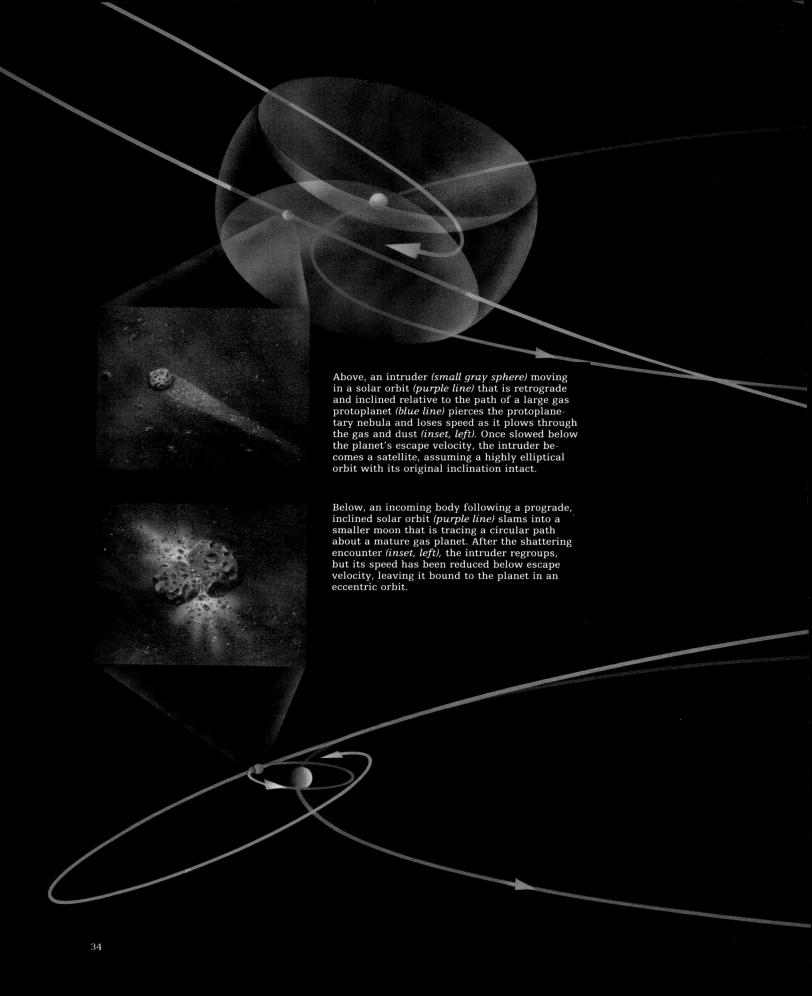

Above, an intruder *(small gray sphere)* moving in a solar orbit *(purple line)* that is retrograde and inclined relative to the path of a large gas protoplanet *(blue line)* pierces the protoplanetary nebula and loses speed as it plows through the gas and dust *(inset, left)*. Once slowed below the planet's escape velocity, the intruder becomes a satellite, assuming a highly elliptical orbit with its original inclination intact.

Below, an incoming body following a prograde, inclined solar orbit *(purple line)* slams into a smaller moon that is tracing a circular path about a mature gas planet. After the shattering encounter *(inset, left)*, the intruder regroups, but its speed has been reduced below escape velocity, leaving it bound to the planet in an eccentric orbit.

SNARING A SPEEDING INTRUDER

Although the capture scenario does not apply to Earth's moon, it appears to explain the origins of many satellites whose paths are inclined to their planet's equatorial plane, highly elliptical, and, in several cases, even retrograde (opposite the planet's direction of spin). Orbits such as these could not have resulted from accretion within a rotating disk, because any moon that formed there would follow a rounded, equatorial, prograde path. Scientists have concluded that such irregular moons were originally passing interplanetary travelers—asteroids or comets—that were drawn into orbit by the planet's gravity.

For a prospective satellite to be captured, it must be slowed below escape velocity—the speed needed to outstrip that gravitational pull. Mission planners who want their probes to orbit a planet achieve this by firing retrorockets as the craft nears its target. But natural forces must account for the deceleration of approaching asteroids or comets. Of several scenarios, scientists regard two as most likely: An incoming body might be slowed by drag as it passes through a protoplanetary nebula *(left, top)* or it could be impeded by a collision with an existing satellite *(bottom)*.

Protoplanetary nebulae provide large targets for passing objects, but they are short-lived, lasting for between a thousand and a million years. So long as the nebula is present, its material will continue to drag at the captured object, causing the moon to spiral into the planet. Thus the majority of satellites captured by this means have likely been swallowed by their gaseous planet, and those that survive today must have been captured just before the nebula dissipated.

Although an existing moon presents a much smaller target than a protoplanetary nebula, collision can take place at any time after satellite formation. At least one new moon will likely form out of the remnants of such a collision, and several are possible.

A Satellite Spawned by Catastrophe

In the first act of the giant impact scenario, a Mars-size body collides with Earth at an oblique angle, knocking the planet askew and producing intense heat and pressure.

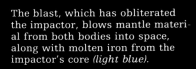

The blast, which has obliterated the impactor, blows mantle material from both bodies into space, along with molten iron from the impactor's core *(light blue)*.

The stream of ejecta begins to orbit the Earth in the direction the impactor was headed. Much of the debris, including the iron, rains back onto Earth.

A few hours after impact, a disk of ejecta forms around the planet, eventually cooling and condensing into discrete particles.

After about 100 years, the disk has expanded beyond the Roche limit, thus reaching the distance at which a satellite can accrete. As the Moon takes shape, it gradually sweeps up the remaining debris in its path.

Until the giant impact theory emerged, scientists knew more about how distant planets obtained satellites than about how Earth acquired its moon. Support for the fission theory dwindled when it became clear that the ejection of material by the proto-Earth required an improbably high spin rate. The capture hypothesis, for its part, seemed ill suited to a small rocky planet like Earth, which could never have possessed the extended gaseous envelope or massive gravity that helped the large outer planets corral satellites. The remaining option—that the Moon accreted near Earth from the same basic materials—was all but ruled out when lunar missions revealed an absence of such common terrestrial volatiles as water on the Moon and an apparent scarcity of iron there. A new theory was needed to account for these odd discrepancies.

Proponents of the giant impact theory did just that using computer models of the collision and its aftermath *(left)*. Around 4.6 billion years ago, they concluded, an object roughly the size of Mars struck the infant Earth a glancing but devastating blow—powerful enough to tip the planet on its side. The heat and pressure of the collision melted the impactor and part of Earth's mantle, vaporizing water and other volatiles and spewing out the remaining materials in a torrid stream of ejecta. Much of that debris, including heavy molten iron from the impactor's core, fell back to Earth. The matter left in orbit soon formed a disk around the planet, and the disk particles eventually accreted to yield the Moon—an arid body whose low density relative to Earth reflected the filtering out of much of the impactor's iron.

of every rock hound on Earth by retrieving the first Moon stones. Those souvenirs—and the other geologic samples gathered by Armstrong and fellow astronaut Edwin "Buzz" Aldrin during their memorable moonwalk—were carried back to Earth in a special isolation container, just in case they harbored alien and potentially harmful life forms. After being quarantined, the samples were subjected to a month-long battery of tests by a team of twenty-six scientists. Over the next three and a half years, five more Apollo landings added to the geologic trove—in all, more than 800 pounds of rock were collected. Soon, scientists around the world would be vouchsafed pieces of the precious stuff and would publish thousands of pages of analysis. The significance of this work was all but lost on the general public, which acknowledged Apollo as a technological coup but wondered if it had lived up to its scientific billing. In fact, the lunar samples were every bit as revealing as the experts had hoped, although it would take some time to piece the evidence together into a coherent story.

In some respects, the findings confirmed expectations. The maria, for example, turned out to be volcanic in origin, as most scientists had anticipated. But the mare basalts appeared to be surprisingly young for a body that some had assumed to be dormant almost from birth: Samples ranged in age from 3.1 to 3.8 billion years, leaving a substantial gap between the lava flows and the period of heaviest impacts, which presumably occurred soon after the Moon's formation 4.6 billion years ago. Evidently, the deep fissures left by the huge impacts provided an avenue for magma that ultimately forced its way to the surface—a conclusion that appeared to demolish Urey's cold-Moon hypothesis. Indeed, there was strong evidence that the Moon had been forged under conditions of extreme heat. Unlike Earth, it was devoid of water either in free form or bound up in rocks and was quite low in other volatiles such as lead and potassium that vaporize at high temperatures.

The relative scarcity of these volatiles and of other common terrestrial elements such as iron threw fresh light on the underlying question of how the Moon came into being. The findings spelled trouble for the coaccretion theory, which assumed a basic chemical kinship between Earth and its satellite but could not explain why water, a major terrestrial component, should be missing from the Moon. An equally damaging blow was delivered to the capture hypothesis by the discovery that the Moon rocks were, in other respects, hauntingly similar to those on Earth. As one planetary scientist put it, "With a little chemical diddling you could say that the moon and earth are from the same parent body." That left only the fission theory, but the evidence accumulating against that hypothesis was compelling. For one thing, it could not account for the fact that the Moon at its inception was apparently hotter than Earth. In short, none of the established origin theories seemed to fit the facts, leading one scientist to joke that because the Moon defied rational explanation, it must not exist.

Yet the tangible evidence of the Moon rocks could not be denied. Even as the dominant theories crumbled, a new hypothesis was emerging that knit together the seemingly contradictory bits of testimony from the lunar missions. Although a number of scientists helped elaborate this fresh account of the Moon's genesis, it was first articulated in 1974 by William Hartmann and Donald Davis of the Planetary Science Institute in Tucson. Hartmann later explained that their efforts had been inspired in part by the work of Soviet astronomers who traced some of the puzzling anomalies of the Solar System to devastating impacts by large planetesimals. In 1966, for example, astronomer Victor Safronov had proposed that the odd orientation of Uranus—which spins on its side—was the result of a collision that literally bowled the planet over. Earth tilts sideways, as well, albeit at a less severe angle of twenty-three degrees. Hartmann and Davis wondered if the home planet, too, had once suffered a wrenching impact that endowed it, by chance, with its uncommon satellite, the only sizable moon in the inner Solar System. They began by estimating the size of the largest planetesimal that could have struck Earth in its formative stage. They knew that the huge basins on the Moon that later filled with lava had been carved roughly four billion years ago by bodies at least ninety miles wide. It followed that earlier on, in the Solar System's tumultuous planet-forming stage, even larger planetesimals would have been careering through space. Hartmann and Davis concluded that a few bodies the size of the present Moon or somewhat bigger could have posed a threat to Earth as it consolidated.

The two scientists then calculated the effects if such a huge planetesimal had hit home. The fierce encounter would have melted much of Earth's mantle and nearly all of the incoming body but its core, sending hot debris spewing into space—and perhaps knocking Earth askew *(page 36)*. Much of the debris would have fallen back to the planet, but the rest would have been lifted high enough to begin orbiting Earth. After about 1,000 years, less than an eye-blink in cosmic terms, the orbiting matter would have amalgamated into a Moon-size object whose low density relative to Earth reflected the fact that it was derived mainly from lighter mantle materials rather than heavier core components. This scenario—known as the giant impact, or big whack, theory—helped explain the central paradox of the recent lunar data: that the Moon was akin to Earth in some respects and alien in others. The peculiar character of the satellite reflected its fiery birth and its mixed ancestry: part Earth, part intruding planetesimal.

Hartmann outlined this scenario with some trepidation at a conference at Cornell University in 1974. He feared that the experts would spot a flaw in the theory and tear it to shreds. After delivering his speech, he braced himself for the worst when Alastair G. W. Cameron of the Smithsonian Astrophysical Observatory in Cambridge, Massachusetts, rose to offer his comments. Cameron was a leading authority on the evolution of planets and their satellites. To Hartmann's surprise, Cameron announced that he and a colleague at the Jet Propulsion Laboratory, William Ward, had been thinking along similar lines. They agreed with Hartmann and Davis that the Moon had been spawned through impact, but they had arrived at a different estimate of the size of the

Assaying the Lunar Samples

Between 1969 and 1976, six Apollo and three Soviet Luna missions carried back to Earth nearly 850 pounds of rock scavenged from various landing sites on the Moon's surface. This considerable collection, examples of which are shown below, afforded geologists a richly detailed view of both the Moon's composition and its history. Unlike Earth's rocks, which contain water, those retrieved from the Moon are bone dry, suggesting that they formed in a waterless environment.

All the lunar samples are igneous—molten rock that cooled and crystallized. Evidently, the young Moon was covered by a magma ocean or was perhaps entirely molten. As the magma cooled, it differentiated; the heavier elements sank and the lighter ones floated upward to form a crust. This ancient crust, more than four billion years old, is now confined to the lunar highlands, or terrae, and consists mainly of anorthosites *(middle example below):* lightweight compounds made up principally of aluminum and calcium. Elsewhere, the Moon's original crust was shattered and recemented by a furious onslaught of impacts that occurred between 4 billion and 3.8 billion years ago. The heat of these impacts produced breccias *(bottom):* mixtures containing fused fragments of original rock types. The largest explosions during the period of intense bombardment blasted out huge basins and left deep fissures beneath them. Magma oozed up through the fissures and, over the course of hundreds of millions of years, coated the big basins, now known as maria, with a veneer of basalts *(top):* darker, heavier rocks than the crustal anorthosites.

Mare basalt. Like other such samples, this chunk of mare basalt contains fragments of material that crystallized out of magma that rose from the depths. A photomicrograph of a polished slice of basalt *(left)* shows plagioclase *(white)*—the main constituent of the crustal sample below—intermingled with darker minerals, including pyroxene *(blue)* and ilmenite *(black)*, both of which contain iron and titanium.

Anorthosite. Retrieved by the astronauts of *Apollo 15,* this sample taken from the lunar highlands was dubbed the "Genesis Rock" for its relative antiquity. The rock is composed largely of plagioclase, which is a light-colored, lightweight mineral that accounts for the highlands' bright appearance relative to the maria. When viewed under a microscope, the anorthosite has a striated look, a function of different orientations of the crystallizing mineral.

Breccia. This sample is a conglomerate of tiny rock fragments trapped in a glassy matrix by the heat and pressure of an explosive impact. In the closeup, vestiges of plagioclase from the pulverized crust show up as white against the dark crystalline base.

impacting body. To account for the present angular momentum of the Earth-Moon system, Cameron ventured, the incoming planetesimal must have been nearly half as wide as Earth, or roughly the size of Mars.

That the Moon resulted from a catastrophic smash-up was an idea that some scientists found unattractive, since it depended on blind chance rather than any predictable or purposeful process. Hartmann, for his part, preferred to think of random collisions not as senseless flukes but as spice in the cosmic stew. "Although the process of planet formation was more or less orderly and gradual," he wrote, "occasional big impacts may have superimposed some variety on an otherwise dull solar system." Whether or not Hartmann's colleagues shared his aesthetic appreciation for such quirks of fate, they were forced to admit that the giant impact accounted for facts that no other origin hypothesis could accommodate. In refining the theory, Cameron and Ward argued persuasively that the intense heat of the collision would have boiled away water and other volatiles, explaining their absence on the Moon. By the mid-1980s, most planetary scientists were inclined to agree that the Moon had been conceived in a torrid encounter between Earth and a celestial intruder.

Like all theories, this one would have to stand the test of time. But it represented a persuasive synthesis of the available lunar data. In a sense, the investigation of the satellite's shadowy past had come full circle. As George Darwin had suggested a century earlier, the Moon was evidently descended from Earth, although the process was more complex and explosive than the simple spinning-off of terrestrial material he had envisioned. In any case, once the material jetted aloft by the giant impact accreted into a sizable body, it behaved much as Darwin described, raising tidal bulges on Earth that served to slow the planet's spin and sent the Moon spiraling slowly outward. In the process, the young satellite encountered other chunks of material left over from the collision, some of them quite large. These moonlets plowed into the satellite, kindling its interior and launching a period of geologic ferment that culminated in the great magma eruptions *(pages 41-45)*. Around three billion years ago, the turmoil subsided. Since then, the Moon's surface has been disturbed only by sporadic meteorite impacts, by minuscule missiles hurtling in from the Sun and cosmos that slowly built up a layer of dust—and, of late, by restless visitors from the parent planet, who drove their stripped-down, mesh-wheeled rovers across the barren landscape in search of the lunar essence. They left their own inscrutable messages, like some updated cuneiform, in the gray powder. With nothing but the shower of micro-meteorites liable to erode them, the footprints and tire tracks inscribed by human curiosity should survive for millions of years.

A LUNAR CHRONICLE

Built from the detritus of a world-ripping collision, shaped by internal fires and the hammer-blows of space debris, the Moon is a fossilized tale of violence. Hints of long-ago tumult can be read in two kinds of lunar terrain visible from Earth: bright highlands (called terrae, the Latin word for "continents") and dark plains (known as maria, Latin for "seas," because they resemble bodies of water). Every square mile of the highlands was savaged during a period of meteorite bombardment that ended about 3.9 billion years ago and left some scars as large as or larger than the state of Texas. The maria, representing about a sixth of the surface area, hide the worst of this punishment. In effect, they are a second skin—outpourings of lava that spread over the largest meteorite-gouged basins.

In recent decades, the Moon has surrendered its secrets to a variety of investigative techniques: spectroscopic and radar studies from Earth; precision analysis of how the lunar mass affects orbiting spacecraft; and, most revealing of all, U.S. and Soviet landings that yielded closeup views and 840 pounds of rock samples. From this evidence, scientists have been able to reconstruct the main steps in the Moon's evolution, a story recounted on the following pages. Lacking the size to maintain its heat, the Moon ceased internal activity roughly 850 million years after the end of the era of heavy bombardment. Earth's companion has been geologically moribund ever since—even as volcanoes, earthquakes, and the shifting of continents have changed the face of the planet itself.

A Molten Infancy

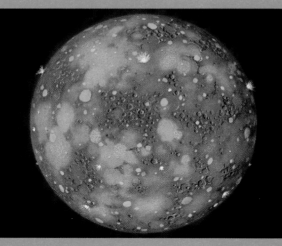

About 4.6 billion years ago, during the same era that saw the planets and Sun coalesce out of a wheeling nebula of dust and gas, Earth's moon was born. Its genesis, as currently envisioned by most lunar scientists *(page 36)*, was traumatic in the extreme. According to this so-called giant impact theory, a Mars-size body smashed into the proto-Earth, sending a great cloud of debris into orbit around the planet. Matter in the cloud quickly cooled and condensed into particles, which subsequently collided and began to accrete. Collisions occurred with such frequency, however, that the heat turned the accreting rock into magma to a depth of hundreds of miles.

The assembly of the Moon did not end its birth pangs. Leftover collision-fragments still orbiting the Earth slammed into the lunar surface, maintaining its molten state *(above)*. In time, the rate of impact slowed, and the surface cooled enough to allow a thin film to form, like the skin on hot cocoa, insulating the magma from the chill of space and allowing it to cool slowly. Minerals crystallized in the torrid depths, rising or sinking according to their density: Relatively dense substances drifted to the bottom of the magma ocean, while those less dense—in particular, a calcium- and aluminum-laden mineral called plagioclase feldspar—floated to the top. By about 4.3 billion years ago, this feldspar froth had solidified to form a crust averaging forty-six miles in thickness.

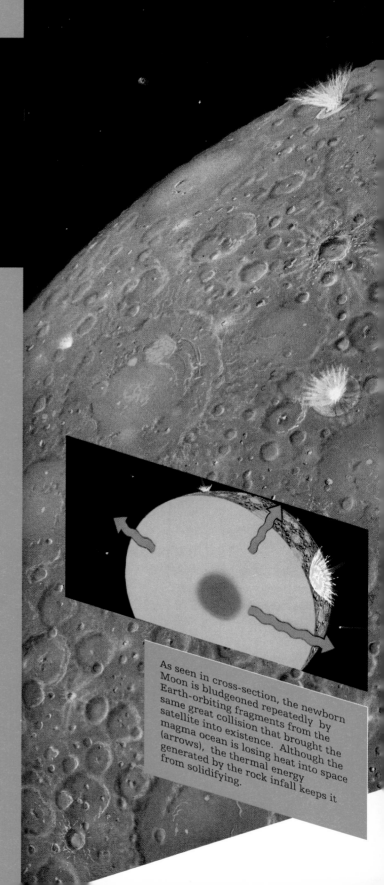

As seen in cross-section, the newborn Moon is bludgeoned repeatedly by Earth-orbiting fragments from the same great collision that brought the satellite into existence. Although the magma ocean is losing heat into space (arrows), the thermal energy generated by the rock infall keeps it from solidifying.

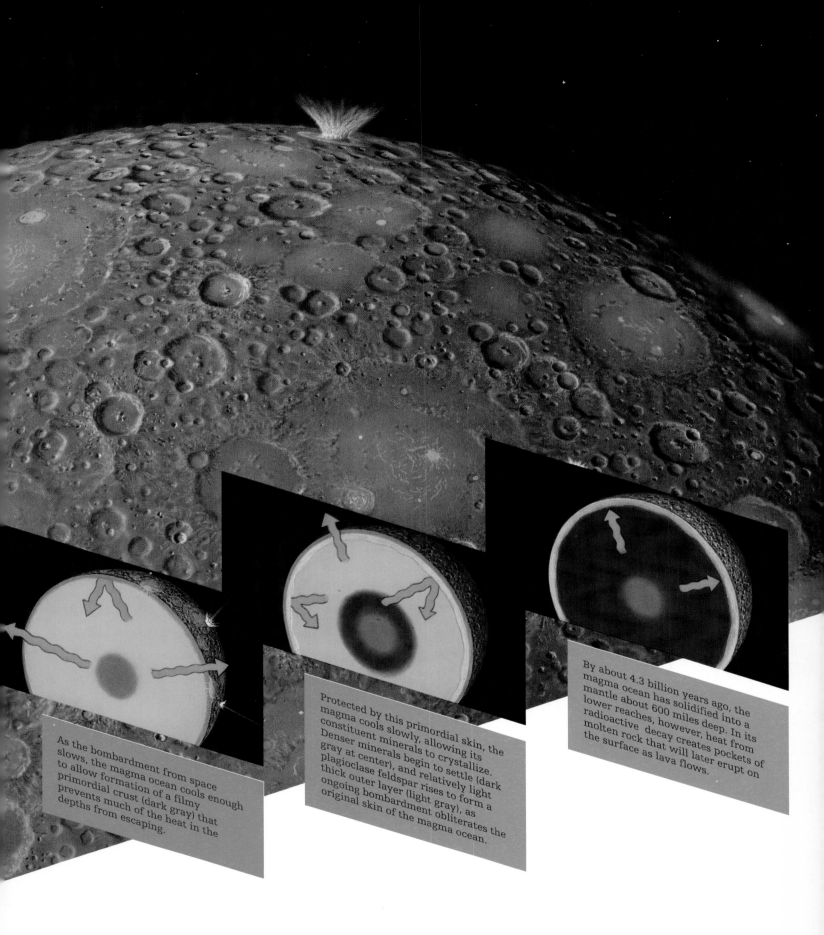

As the bombardment from space slows, the magma ocean cools enough to allow formation of a filmy primordial crust (dark gray) that prevents much of the heat in the depths from escaping.

Protected by this primordial skin, the magma cools slowly, allowing its constituent minerals to crystallize. Denser minerals begin to settle (dark gray at center), and relatively light plagioclase feldspar rises to form a thick outer layer (light gray), as ongoing bombardment obliterates the original skin of the magma ocean.

By about 4.3 billion years ago, the magma ocean has solidified into a mantle about 600 miles deep. In its lower reaches, however, heat from radioactive decay creates pockets of molten rock that will later erupt on the surface as lava flows.

THE ERA OF COSMIC BOMBARDMENT

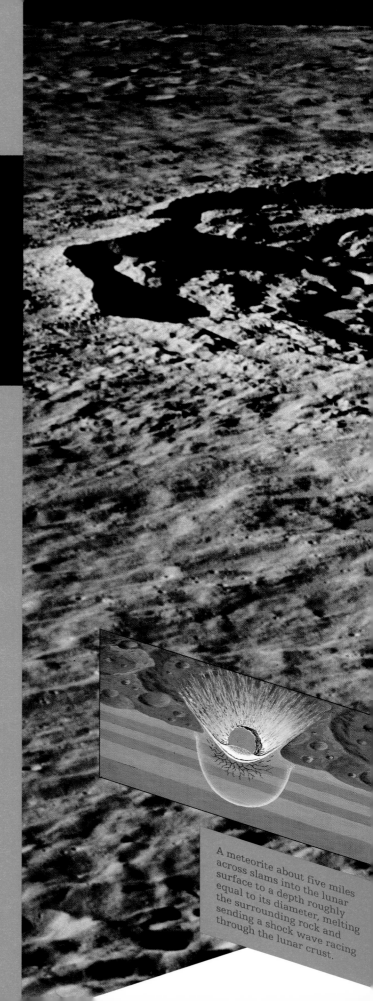

Although debris in Earth orbit accounted for the heaviest barrage of the Moon in its early existence, fragments left over from the more general process of the Solar System's formation contributed to the bombardment, continuing for about 600 million years. By the time the deluge from space abated some 3.9 billion years ago, the Moon was severely scarred *(above).* So, too, was Earth, but weather and geologic activity would erase most of the terrestrial damage.

Even small meteorites could hit with prodigious force. Hurled against the surface at a speed of 10 miles per second, a projectile only 40 feet in diameter would release the energy of a modest-size atomic weapon and excavate a bowl-shaped crater as much as 5 miles across. But many of the intruders were far larger. The Copernicus crater *(right)*—58 miles across—was gouged by a meteorite that measured about 5 miles in diameter and packed a wallop equivalent to 70 trillion tons of TNT, more energy than is contained in the world's entire nuclear arsenal. And even this incredible blast was dwarfed by meteorite blows that carved the largest lunar craters, known as impact basins. Some 3.9 billion years ago, a fast-moving body the size of Rhode Island struck with the fury of billions of H-bombs. Had it been only a little bit larger, scientists believe, it would have blown the Moon to pieces. Instead, it excavated a huge basin ringed by concentric mountain chains, a feature of very large impact craters; the outermost ring measures nearly 750 miles across. The entire event lasted about two hours.

A meteorite about five miles across slams into the lunar surface to a depth roughly equal to its diameter, melting the surrounding rock and sending a shock wave racing through the lunar crust.

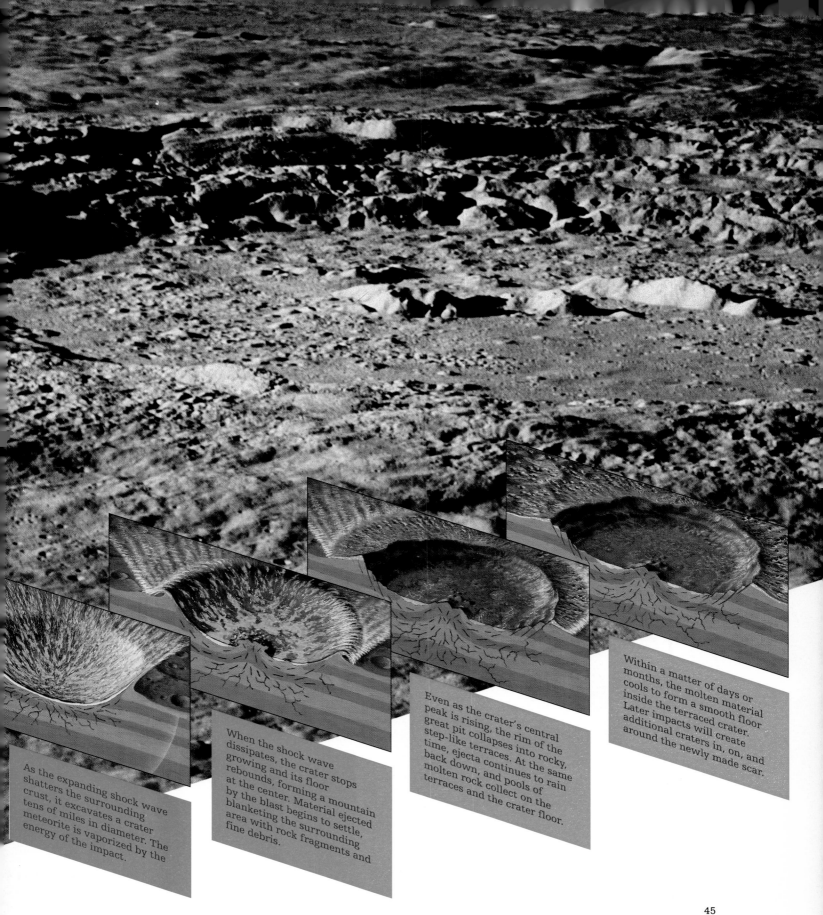

As the expanding shock wave shatters the surrounding crust, it excavates a crater tens of miles in diameter. The meteorite is vaporized by the energy of the impact.

When the shock wave dissipates, the crater stops growing and its floor rebounds, forming a mountain at the center. Material ejected by the blast begins to settle, blanketing the surrounding area with rock fragments and fine debris.

Even as the crater's central peak is rising, the rim of the great pit collapses into rocky, step-like terraces. At the same time, ejecta continues to rain back down, and pools of molten rock collect on the terraces and the crater floor.

Within a matter of days or months, the molten material cools to form a smooth floor inside the terraced crater. Later impacts will create additional craters in, on, and around the newly made scar.

ADDING THE FINAL TOUCHES

Even as great chunks of rock pounded the Moon, other forces stirred the young world. Far below the surface, intense heat from the decay of radioactive elements recreated pockets of magma in the newly solidified mantle. These reservoirs of molten rock found an outlet in impact basins where the most violent meteorite blows had cracked the crust. Magma welled up along the fractures, filling the depressions with fiery seas. Slowly, over a span of 700 million years, repeated outpourings and cooling of the lava formed the maria: vast expanses of basalt, a volcanic rock found on Earth in such places as Hawaii and Iceland. By about 3 billion years ago, the creation of maria had thoroughly transformed the face of the Moon *(above)*. Thereafter, as the Moon's internal heat dissipated, upwellings of lava became a rare event.

In the eons since the time of lava seas, the most significant force shaping the Moon's surface has been the continued fall of modest-size meteorites and a steady rain of grain-size ones. Because of the ceaseless pulverizing and churning of surface rock by these lesser meteorites *(bottom right),* the Moon today is covered with a thick layer of dust and rock fragments, known as regolith. Much shallower on the maria, the regolith is elsewhere as much as forty-five feet deep. The astronauts who walked on this alien soil found that it resembles powdered graphite. The sharply chiseled footprints they left behind *(right)* will endure for millions of years before the slow but inexorable sandblasting from space erases them.

46

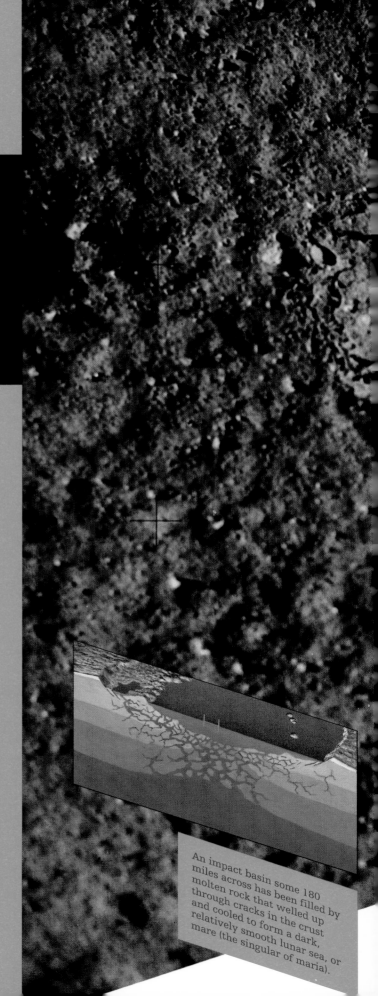

An impact basin some 180 miles across has been filled by molten rock that welled up through cracks in the crust and cooled to form a dark, relatively smooth lunar sea, or mare (the singular of maria).

In this closeup of a 100-foot section of the mare, a meteorite lands between the scars of two previous impacts. Debris (yellow) will fall around the new crater in a thin layer called an ejecta blanket.

The impact creates a bowl-shaped crater whose ejecta blanket (light gray) covers the deposits from the smaller craters on either side.

Tens of millions of years later, the once-smooth mare is pocked by impacts. Ejecta blankets have covered the original surface with a layer of dust, rock fragments, and bits of glass formed by the sudden cooling of molten debris. A continual rain of small meteorites grinds the rocks into dust to produce one-sixteenth of an inch of soil every one million years.

After 100 million years, the debris layer is so thick that only the deepest impacts excavate the original lava surface. Most simply redistribute the upper layer of dust and rock fragments.

ABIDING MYSTERIES

Although more than forty lunar missions, including those carried out by robotic probes, have made the Moon the best-known world beyond Earth itself, some fundamental mysteries remain. The most compelling came to light when the first views of the far side of the Moon were sent back to Earth by the Soviet craft *Luna 3* in October 1959. To scientists' great surprise, considerably less of that formerly hidden lunar hemisphere *(top)* is covered by dark maria than is the case on the side facing Earth *(bottom)*. Indeed, any maria visible are actually features right on the border between the near and far sides.

Several possible explanations have been advanced, most of them building on still-unverified hints that the crust is thicker on the far side and thus less likely to have been penetrated by magma welling up from below the surface. One hypothesis holds that a monstrous impact early in the history of the Moon could have caused crustal variations. Perhaps the Oceanus Procellarum—a mare that is 1,500 miles across (visible in the upper left quadrant of the Earth-facing side shown here)—is the remnant of a blow so mighty that large quantities of crustal material were displaced. Another possibility, a variation on the giant impact theory of lunar formation, is that the Moon was formed by the collision of two smaller bodies that accreted after the initial impact. If the submoons were geologically distinct, the product of their union might not have a uniform crust.

Even if the Moon's crust does not vary in thickness, the two-moon hypothesis might account for an asymmetry in maria. If one of the submoons contained more radioactive material than the other, heat-driven geologic activity might have been greater on one side of their conjoined body. For the moment, however, scientists can only make guesses. The solution must await further lunar exploration and more complete knowledge of what lies below the surface of Earth's singular satellite.

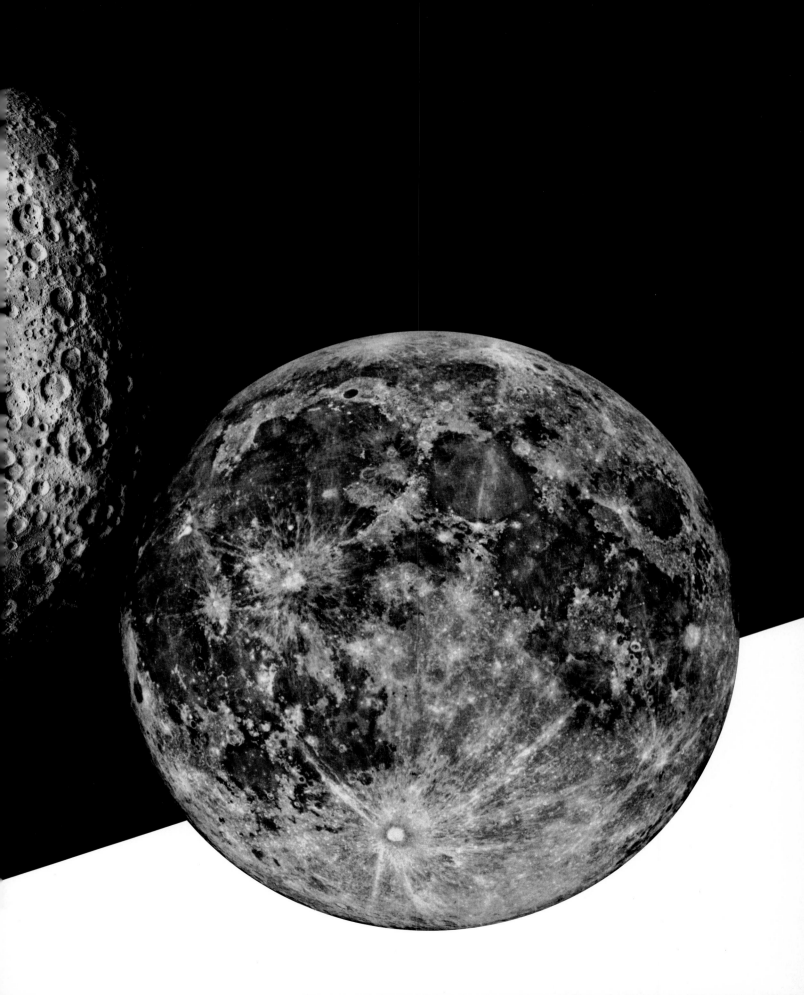

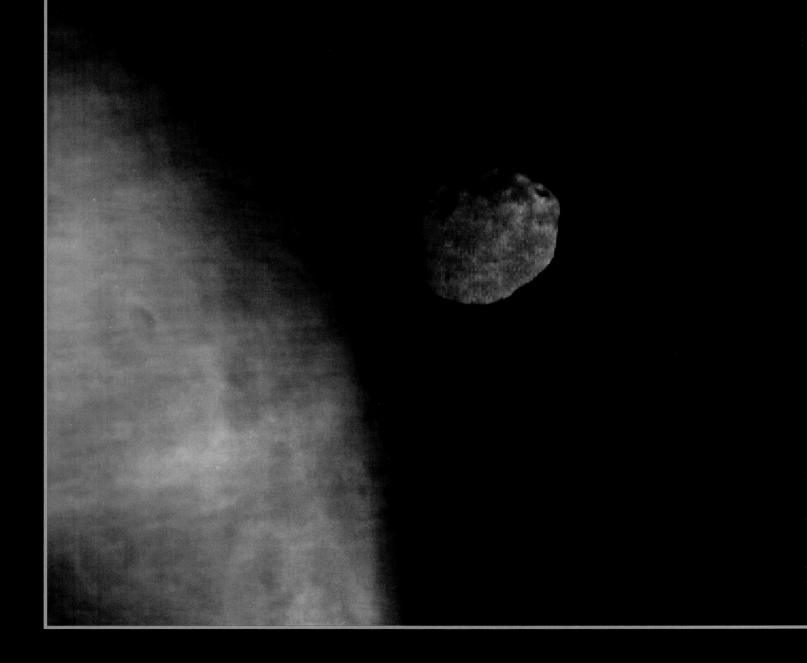

The individual features of Phobos, a small, misshapen rock orbiting Mars *(far left)*, and the planet-size Jovian satellites Io *(below, foreground)* and Europa *(below, right)* are revealed by spacefaring probes that have given astronomers close-up views of moons that were once mere specks of light seen through earthbound telescopes.

During the summer of 1877, Asaph Hall of the U.S. Naval Observatory in Washington, D.C., decided to go hunting for hidden moons around Mars. Like James Christy, who would discover the moon Charon orbiting Pluto by examining photographs at the Naval Observatory a century later, Hall was fixing his sights on a planet that most authorities considered to be a solitary body, unencumbered by satellites. To be sure, Johannes Kepler had long before posited the existence of two Martian moons, but for reasons that were more philosophical than scientific. Seeking signs of cosmic harmony, Kepler had suggested in 1610 that the number of moons orbiting each planet beyond Earth should increase by precise intervals, with the predicted Martian pair representing a neat step between Earth's single satellite and the four known to exist around Jupiter. Although unsupported by evidence, this idea apparently struck a chord. In 1726, satirist Jonathan Swift echoed it in the novel *Gulliver's Travels,* which told of a people called the Laputans who lived on a floating island in the sky and profited by that vantage point to spot "two lesser stars, or 'satellites,' which revolve about Mars." Such fictional feats aside, earthbound observers could find no trace of the Martian moons. The celebrated astronomer William Herschel looked for them without success in 1783, and for generations afterward, few others even bothered to try. By the time Asaph Hall took up the challenge, the astronomy textbooks said flatly that there was nothing to be found around Mars.

Hall was not one to defer to expert opinion when he could make his own determinations. Through sheer will power, he had elevated himself from day-laboring carpenter to one of the nation's top astronomers. He had paid for his education with his carpenter's wages, married his college mathematics teacher—Angeline Stickney—and begun his second career as a lowly staffer at the Harvard Observatory in Cambridge, Massachusetts. In spare moments Angeline taught her husband German, enabling him to keep up with European physics and astronomy. But while culling all he could from the texts, he continued to trust first in his own instincts and vision. For example, while scrutinizing Saturn and its satellites in December 1876, he found that the official estimates of the planet's rotation rate were about fifteen minutes off, and he confidently corrected them. The experience led him to suspect that the authorities might be wrong about the moons of Mars, too.

Hall approached his task with more than a little urgency. Mars was coming into opposition that August, at which time Mars and the Sun would line up on opposite sides of Earth, thus bringing the two planets close together. Such an alignment was known to occur once every twenty-six months, providing excellent viewing on the nights surrounding the event, but due to differences in the shape of the orbits of Mars and Earth, some oppositions brought the planets nearer than others. This would be the closest in fifteen years—a golden opportunity for an astronomer with a first-class telescope. The Naval Observatory's new twenty-six-inch refractor was the largest in the world. But Hall had recently learned that a twenty-seven-inch refractor was being built in Vienna, Austria, and the news strengthened his resolve to search for Martian satellites before competitors with a keener instrument claimed the honor of spotting them first.

PENETRATING THE FOG

Hall began his moon hunt in early August, working late into the night at the Naval Observatory, which was then located on the banks of the Potomac River, near the center of Washington. He searched near the planet's rim because theory suggested that any satellites small enough to have gone undetected must be quite close to Mars; otherwise they would have been pulled away by the Sun's gravity. After scouring the darkness a short distance from the planet, Hall moved his scope in until Mars was nearly in view and tried to discern starlike objects in the surrounding glare. It made for difficult viewing, and after several lengthy sessions, Hall still had nothing to show for his pains. His wife urged him on, however. On the evening of August 11, she packed him off to work as she had on the evenings before—with a prepared meal and words of encouragement. At half past two the next morning, Hall spotted a faint object just north of the planet; but no sooner had he fixed its position than fog rising from the river spread over the observatory and ended the session. For several days, fog and clouds continued to blanket the area, leaving Hall in suspense as to whether he had detected a Martian satellite or merely a background star. At last, on August 16, he caught a clear view of the same faint object near Mars, moving along with it in the unmistakable pattern of a moon bound by the planet's gravity.

The next night, Hall spotted another faint object even closer to the planet. As he followed this second elusive satellite, occasionally losing track of it, he found to his surprise that it appeared on different sides of Mars in the course of a single night. No moon was known to orbit so swiftly about its primary, and Hall wondered if he was actually observing two or three tiny satellites circling Mars at roughly the same distance and popping in and out of view. Yet it soon became evident, he wrote later, "that there was in fact but one inner moon which made its revolution around the primary in less than one-third the time of the primary's rotation—a case unique in our solar system." Hall chose evocative names for his twin finds. He named the swift inner satellite Phobos and the outer one Deimos, after the legendary Greek figures

Fear and Terror who attended the war god Mars.

After announcing his discoveries, Hall went on to determine the moons' orbits more precisely. Both traced nearly circular paths about the planet's equator, it turned out, with Deimos located about 12,500 miles from the surface of Mars and Phobos a mere 3,760 miles from the planet. Hall fixed the period of Deimos at thirty hours and eighteen minutes and that of Phobos at just seven hours and thirty-nine minutes; thus, a hypothetical observer standing on Mars would see its swift inner moon rise and set at least twice a day while the outer moon, lagging behind the planet's rotation period of twenty-four hours and thirty-seven minutes, would appear to move slowly in the opposite direction, dawdling in the sky for more than two days before receding below the horizon. Orbital peculiarities aside, the two satellites were remarkably small. Early calculations based on the amount of light they reflected com-

pared to Mars put them both at a diameter of six or seven miles. Later observations increased those estimates somewhat, but Phobos and Deimos were plainly tiny when compared to Earth's moon or the four major satellites of Jupiter identified by Galileo.

Ostensibly, the detection of the two Martian moons bore out Kepler's time-worn prediction, but the find did little to support his assumption that the planets and their satellites conformed to some neat design. Indeed, the very existence of the two small orbiters raised the possibility that other strange minimoons might be lurking around the outer planets. In the early 1900s, astronomers would spot several such satellites around Jupiter, following highly eccentric orbits that brought a new level of complexity to the once-tidy picture of the Sun's family. But not until rocket-launched probes began their outward journeys of discovery later in the twentieth century would planetary scientists fully appreciate the Solar System's astonishing lunar diversity. The probes would reveal a crazy quilt of surfaces covering the Jovian moons alone—some of them ancient and dark, others gleaming with fresh coatings of ice, still others rent by volcanic outbursts. Analyzing the hectic pattern, astronomers would begin to decipher the cryptic process of celestial evolution that separates satellites into distinct species and distinguishes in turn among the members of each group.

A SATELLITE ON THE FAST TRACK
Faint and diminutive, the Martian moons Phobos and Deimos yielded little information to astronomers for well over half a century following their discovery. Then in 1945, B. P. Sharpless of the Naval Observatory made a sur-

In August of 1877, astronomer Asaph Hall *(far left),* shown with a globe of Mars, discovered the Martian satellites Deimos and Phobos using a twenty-six-inch refractor at the United States Naval Observatory *(left).* After searching nightly over the course of some two weeks, Hall spotted the tiny moons, whose close-in orbits had kept them hidden in the planet's reflective glare.

prising find as he scrutinized the orbit of Phobos. Comparing his own observations to those done decades earlier, he determined that the moon was drawing nearer to Mars at the rate of about five yards a century and speeding up ever so slightly in the process, the inevitable result when a satellite moves into a tighter orbit around its primary. Sharpless estimated that if this process, known as secular acceleration, continued at the present rate, Phobos would crash into Mars in about 100 million years. Later calculations lowered that estimate to a range of 40 to 70 million years, depending on whether the moon remains intact until it collides with Mars or is torn apart earlier by tidal interactions with the planet—an outcome that might endow Mars with a ring of satellite fragments.

Astronomers offered various explanations for this peculiar orbital evolution of Phobos. The most intriguing hypothesis was put forward in 1959 by the Soviet astronomer Iosef S. Shklovsky, who began with the assumption that the atmosphere of Mars must be dragging the small moon down into its ever-tighter orbit. Shklovsky acknowledged that the Martian atmosphere was too thin to cause the observed secular acceleration of Phobos if it was a solid body. But atmospheric drag might indeed be the explanation if the moon was hollow—which implied that Phobos was not a natural satellite but an artificial one, fashioned by intelligent creatures. Perhaps, Shklovsky ventured, Phobos had been "launched into orbit in the heyday of a technical civilization on Mars, some hundreds of millions of years ago."

Many scientists were reluctant to take this suggestion seriously. But the questions Shklovsky addressed could only be answered with more detailed views of Phobos and Deimos, which remained mere specks of light through the lenses of the most powerful telescopes. The veil of mystery shrouding the two moons at last began to lift in 1969 when the American spacecraft *Mariner 7*—one in a series of probes investigating Mercury, Venus, and Mars—took a passing snapshot of Phobos silhouetted against the bright disk of Mars. This first closeup revealed the moon to be irregular in shape, quite dark, and about twelve miles wide. Two years later, after *Mariner 8* had been destroyed in a launch accident, *Mariner 9* came along. This scout was programmed to orbit Mars and focus on its surface. But as the probe drew near in November 1971, the surface of Mars suddenly vanished in a planetwide dust storm. Few disturbances of this magnitude had ever been observed before. Only the tops of ancient volcanic mountains protruded above the clouds. Valuable time might have been lost—for the clouds stayed in place until December—had not astronomer Carl Sagan prevailed on mission planners to alter the craft's course and bring it closer to the small moons. Thanks to these efforts, *Mariner 9* obtained spectacular views of Phobos and Deimos while Mars remained shrouded in dust.

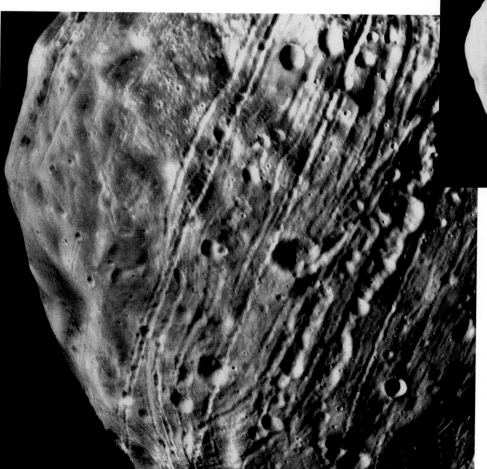

Phobos. A prominent scar on the face of the larger of the two Martian satellites, six-mile-wide Stickney crater shows deep grooves *(above and left)* that appear to be rows of small pits. Phobos is no more heavily cratered than its sister Deimos *(right),* but its relatively thin blanket of regolith exposes more of its battered surface.

The spacecraft's camera showed Phobos to be a lopsided pile of rock—or, as Sagan described it, "a diseased potato." This was no relic of an ancient civilization: Phobos was a natural body, and quite a bleak one at that. On close inspection, in fact, Phobos looked positively decrepit, its great age reflected in the many craters marring its surface. Only a body that had been knocking about the Solar System for billions of years could have accumulated so many scars. The same appeared to be the case with the satellite's slightly smaller but equally disfigured partner, Deimos. The density of craters on both lumpy moons was so high that every new impact would likely remove the traces of an older one. In addition, both satellites were covered with a layer of dusty debris, or regolith, a product of the same sort of pulverizing impacts and incessant micrometeorite bombardment that had coated Earth's moon with a powdery blanket. The presence of regolith on the tiny Martian satellites was something of an enigma. Astronomers wondered how such small bodies could possess enough gravity to retain the debris kicked up by repeated impacts. Some speculated that the ejected

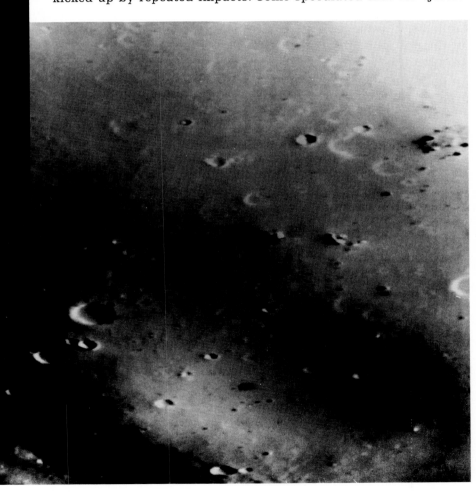

Deimos. Although it appears relatively smooth from a distance *(above)*, a closer look at Mars's outer moon *(right)* reveals dense cratering that is partially obscured by a thick layer of dust. Both Deimos and Phobos are ellipsoidal in shape and are aligned by tidal forces to orbit with their long axes pointed toward the planet.

debris was actually held in orbit near the moons by the gravity of Mars, and that Phobos and Deimos merely swept the material back up as they passed through it. Others suggested that as powder accumulated on the surface, it cushioned the blow of incoming objects, so that little of the regolith was agitated to escape velocity.

In the wake of the *Mariner 9* encounter, a group of cartographers used the data to construct a map of Phobos, showing 260 major craters in three dimensions. The largest, named Stickney in honor of Asaph Hall's wife, measured some six miles in diameter—nearly half the width of the entire moon. The vast cavity had been carved by an impact equivalent to the blast of a multimegaton bomb that must have come close to tearing Phobos apart, an ever-present danger for a moon so slight it lacks the self-gravity even to tug its rocky body into a spherical shape. The pictures from *Mariner 9* also confirmed what astronomers had suspected: Phobos and Deimos are among the darkest objects in the Solar System, reflecting only about six percent of the incident light. Scientists hoped to analyze the faint light to determine the chemical composition of the soil, but the spectrometer on board the probe was not fit for the task, so the problem would have to be attacked by a different method on a later trip. One thing was clear, however: These were the only moons to be found around Mars.

The revelations of *Mariner 9* were rivaled by the disclosures of the next generation of American spacecraft to invade Martian airspace: the Viking orbiters. *Viking 1* flew by Phobos in February 1977 and again three months later, coming within sixty-two miles of the small moon and obtaining precise images that distinguished surface details smaller than fifty yards across. That October, *Viking 2* passed within fifteen miles of Deimos and relayed splendid views of the outer moon, resolving details only a few yards across. In the closeups transmitted by *Viking 1,* Phobos's giant crater Stickney stood out sharply, affording scientists a look at curious features not previously detected—long parallel grooves emanating from the depression. Measuring perhaps twenty yards deep and more than a hundred yards wide, these grooves extended up to twelve miles, curving around the surface of the moon. Scientists speculated that these were deep fractures caused by the collision that forged Stickney.

High-resolution photos revealed a further surprise—the grooves were marked by a series of nearly continuous small craters, pocking the channels like the holes in a flute. Three planetary scientists at Cornell University— Peter Thomas, Joseph Veverka, and Arthur Bloom—theorized that the pits formed when regolith drained into the fractures. This idea was later supported by Kevin C. Horstman and H. Jay Melosh of the University of Arizona, who conducted an experiment in which they covered two contiguous plates of glass with granular material, then moved the plates apart slightly. As the grains poured into the resulting gap, a row of tiny pits much like the groove craters on Phobos appeared atop the sand in the trough. Another possibility, based more on theory than observation, is that liquid deep within the moon turned to gas at some time in the past, creating vents as the vapor forced its way through the fissures and disturbed the dust at the surface.

Deimos, for its part, appeared free of such grooves in the images sent by *Viking 2*. Its largest visible crater measured barely a mile and a half across—the result of an impact that packed too little punch, apparently, to produce long, deep fractures. Overall, Deimos appeared to be smoother than its larger sibling, not because it was any less densely cratered but because regolith had filled in the depressions more thoroughly there than on Phobos, another indication that the abundance of regolith on a moon does not depend simply on the strength of its gravity.

One of the most important disclosures of the Viking missions resulted from a disarmingly simple maneuver. By guiding *Viking 1* into a close encounter with Phobos and plotting the change in the spacecraft's trajectory as it passed by, scientists were able to determine the force of the moon's gravity and, from this, its mass and density. The density reading was remarkably low: slightly more than two grams per cubic centimeter, or only about half that of Mars. The moon's low density, combined with its small size, has endowed it with so little gravity that a person standing on the surface could easily throw a rock into orbit around the moon by tossing it straight ahead at a mere twenty-eight feet per second—less than one-third the speed of a major-league fastball. The rock would circle Phobos and return to the pitcher's mound, as it were, in just an hour and a half.

FUGITIVES FROM THE MAIN BELT?
The lightweight composition and extremely dark surface of Phobos and Deimos suggested to scientists that the moons are made up of the porous black material characteristic of the so-called C-type asteroids—carbonaceous chondrites. Although most such bodies have long been confined to the outer half of the main asteroid belt, between Mars and Jupiter, collisions and other perturbing factors have forced a few of them into highly irregular orbits that periodically intersect the path of Mars. Conceivably, Phobos and Deimos could have been errant asteroids that came close to colliding with Mars but were captured instead, a history that would help explain why they bear so little resemblance to the planet, which is both denser and considerably brighter than its satellites.

In the aftermath of the Viking missions, astronomers partial to this capture hypothesis drew up a scenario explaining how Phobos and Deimos might have been snared by Mars and drawn into their present, regular orbits. In the early phase of the Solar System's evolution, they argued, Mars could have possessed a thick and widely extended atmosphere, whose friction presumably slowed down the passing asteroids enough for the planet's gravity to hold them in orbit. Initially, the captured bodies would have followed pronounced elliptical paths around the planet, but over hundreds of millions of years, tidal forces could have circularized their orbits *(pages 60-61)*. Similar tidal effects might have tugged the two small moons from their original inclined

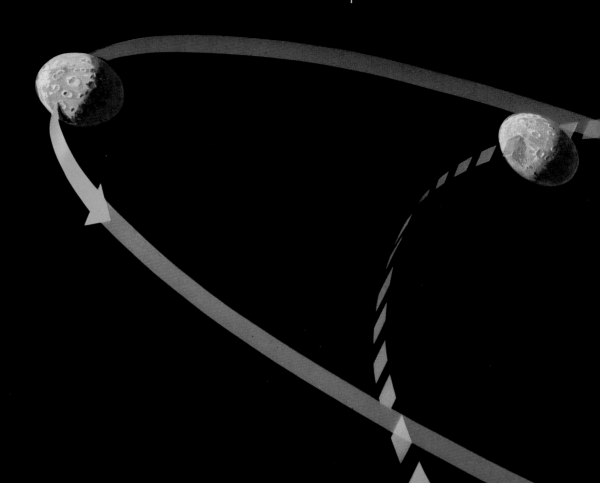

TIDAL ADJUSTMENTS TO ORBITAL ECCENTRICITY

Although many moons in the Solar System travel in nearly circular orbits roughly along their planets' equators and are believed to have been formed by the same gravitational processes that created the planets, others follow looping, eccentric paths tilted to the equatorial plane. These exceptions to the general rule tend to resemble asteroids in that they are irregularly shaped and covered with very dark surface material thought to be carbonaceous; some scientists argue that the satellites are in fact captured bits of debris leftover from the birth of the Solar System. But a few of these asteroid-moons, such as Mars's Phobos and Deimos, are exceptions to the exception, following nearly circular paths that are similar to those of moons presumed to have been created along with the original planetary system.

One explanation for the incongruous orbits, illustrated here, is that such regularized paths are the result of millions of years of tidal pulling. Initially, a captured asteroid takes up an eccentric orbit that subjects it repeatedly to varying gravitational force as it swings close to the planet and then away again. The resulting tides create bulges on the moon's surface that in turn are acted upon by the planet's gravitational pull in such a way as to cause the moon's orbit eventually to become circular.

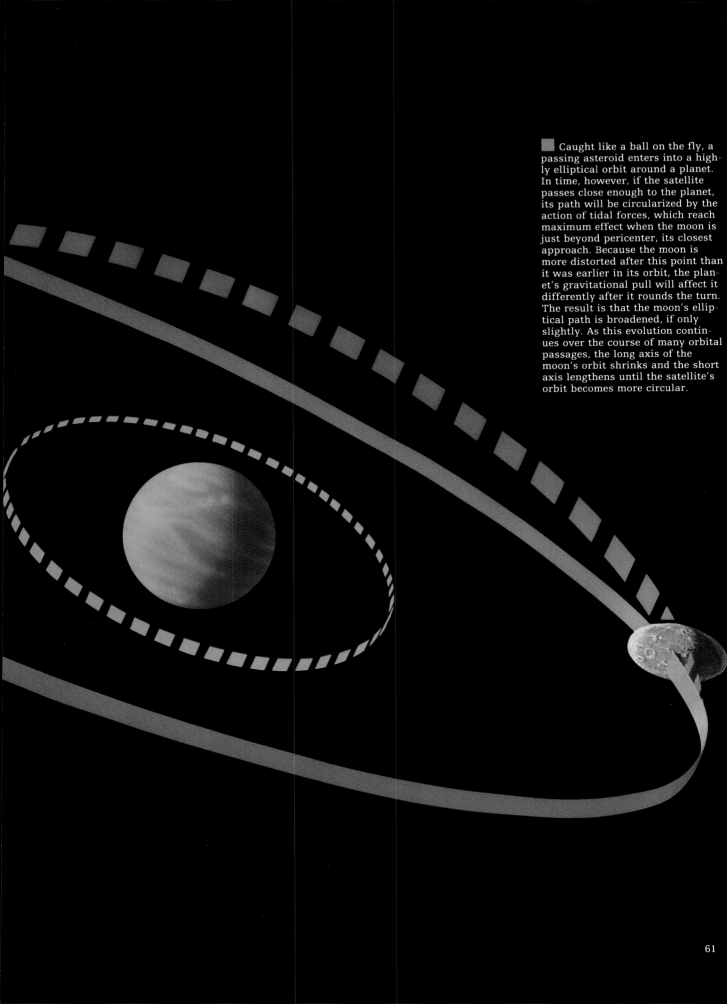

Caught like a ball on the fly, a passing asteroid enters into a highly elliptical orbit around a planet. In time, however, if the satellite passes close enough to the planet, its path will be circularized by the action of tidal forces, which reach maximum effect when the moon is just beyond pericenter, its closest approach. Because the moon is more distorted after this point than it was earlier in its orbit, the planet's gravitational pull will affect it differently after it rounds the turn. The result is that the moon's elliptical path is broadened, if only slightly. As this evolution continues over the course of many orbital passages, the long axis of the moon's orbit shrinks and the short axis lengthens until the satellite's orbit becomes more circular.

Continuing Orbital Evolution

Whatever a moon's origins, its orbit will continue to evolve as a result of gravitational interaction with its primary, as each body tries to maintain its own orbital momentum against the other's gravitational pull. One stage of this process, known as tidal locking or synchronous spin, is common in the Solar System and evident in the relationship between Earth and its moon. In such systems, the planet's tidal pull slows the satellite's rotation until the moon spins once on its axis in the time it takes to complete one orbit of its primary. Thereafter, the planet locks onto the long axis of the distorted moon, causing it always to keep the same face toward the planet.

As evolution continues, moon and planet can achieve the condition called mutual synchronous spin, whereby the moon will complete one revolution around the planet in the same time it takes the planet to rotate once on its axis. As a result, not only will the moon keep one face toward the planet, but the planet will show only one face to the moon. The moon thus will be visible from only one side of the planet and,

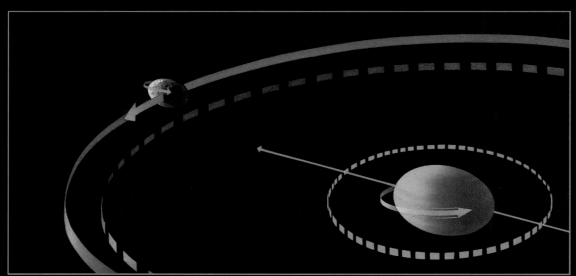

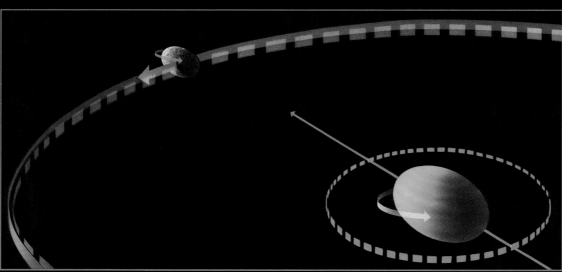

A moon lying just outside the synchronous orbit distance takes slightly longer to complete a full orbit than it takes the planet to spin once on its axis, so the moon lags slightly behind the planet's tidal bulges *(blue arrow, top left)*. As the planet rotates, it exerts a forward pull on the moon, boosting its energy and kicking it into a higher orbit. At the same time, the moon's gravity exerts a backward pull on the planetary bulge, slowing the planet's rotation. After billions of years, moon and planet achieve synchrony *(bottom left)*. The moon orbits at the exact distance that allows its orbital period to match the planet's rotation rate—a state that Earth's moon could reach billions of years from now, when its orbital period and the planet's day would each equal about forty-seven current Earth days.

neither rising nor setting, will hang motionless in the sky—as does Charon in the skies above Pluto.

Whether a moon and planet ultimately reach this stable situation depends on the moon's being able to slow the planet's rate of rotation. As illustrated here, two fates are possible, according to whether the moon is outside the boundary called the synchronous orbit distance. This distance is unique for each planet and is determined by the rate at which the planet spins on its axis. However, because this rate may be slowed over time by the tidal pull of a moon (as well as by gravitational interaction with the Sun), the synchronous orbit distance is subject to change.

A moon that lies within the synchronous orbit distance may be pulled in so close to the planet that it risks destruction, a process that could bring Phobos—now circling Mars at a distance of only 3,760 miles—crashing into the Martian landscape in 70 million years or so. A moon that is just outside the critical distance may be boosted into ever-higher orbits, much as Earth's moon, the Martian moon Deimos, and most of the larger satellites in the Solar System are being pushed outward from their primaries. Many eons hence, such a moon's orbital altitude would coincide with the synchronous orbit distance and it would have achieved ultimate stability.

In this instance, a moon lies well within the synchronous orbit distance, and its orbital speed carries it around the planet in less time than it takes the planet to complete one rotation. The moon thus leads the planet's tidal bulges (*blue arrow, top right*), and the greater pull of the planet acts as a drag on the moon's forward motion. The resulting loss of energy drops the moon into ever-lower orbits. Eventually, the drag brings the moon inside the Roche limit, where intense tidal forces could, in some cases, pull the satellite to pieces. Even if a moon has enough internal cohesiveness to survive inside the Roche limit, however, the victory is only temporary: Some time later, the moon will shatter on the planet's surface, as Phobos is expected to do tens of millions of years in the future.

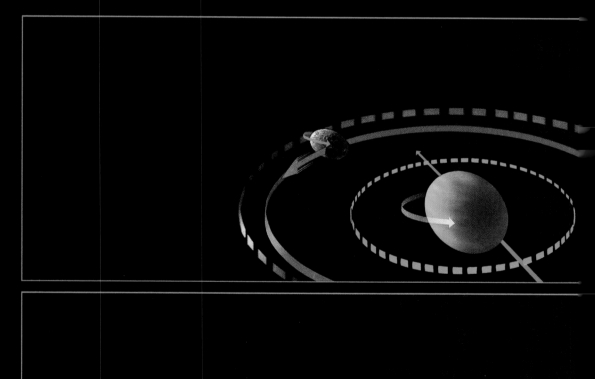

paths toward Mars's equatorial plane.

More difficult for the theorists to explain was why the orbits of two satellites of similar origins should have evolved in contrasting ways. Phobos, as B. P. Sharpless found in 1945, is slowly spiraling in toward Mars, tugged by the planet's gravity into ever-tighter circles *(pages 62-63)*. Deimos, for its part, revolving more slowly about Mars than the planet is spinning on its axis, is being impelled gradually outward like a noose at the end of a swiftly twirled lasso. Ultimately, Deimos may share the predicted fate of Earth's moon and orbit in total synchrony with the planet, locked above one spot on Mars. Attempts to extrapolate the divergent orbits of Phobos and Deimos back in time indicate that they should have met at one point, raising the question of how they could have survived such an encounter. Perhaps they were once part of a single fragile object that was captured by Mars and then sundered by an impact, yielding two fragments, one inbound and the other outbound.

Some astronomers dismiss any such capture scenario as improbable. But the obvious alternative—that the moons of Mars formed through a process of accretion around the parent planet—would account only for their circular, equatorial orbits, while leaving unexplained such anomalies as their peculiarly dark surfaces and their differing orbital destinies. The question of how the satellites formed may not be resolved until astronomers have the same sort of evidence that shed light on the origins of Earth's moon: samples of the surface. Presumably, moons that accreted near Mars would display certain chemical similarities to the parent planet, while captured asteroids would be markedly different.

Scientists hoped to obtain such evidence in 1988 when the Soviet Union launched two new space probes toward Mars. Called *Phobos 1* and *2,* each robot was expected to photograph Mars and Phobos and to conduct various surveys with lasers and other sensors, and then to eject a small landing craft that would touch down on Phobos, sample its soil, and transmit data back to Earth. The probes were launched successfully on July 7 and 12, 1988. But before long, both ran into trouble. On August 28, a flight control officer sent a wrong command to *Phobos 1* as it was making its way to Mars, switching off the thrusters that kept the craft's solar panels oriented toward the Sun. The probe soon lost power and was never heard from again. *Phobos 2* succeeded in traveling all the way to the planet, only to suffer a similar fate as a result of an undetermined malfunction.

A page from the notebooks of Galileo Galilei *(above)* heralds the discovery of four wandering "stars" orbiting in a plane around Jupiter. After tracking the moons' movements, shown in the series of modern photographs at right, Galileo recognized that the stars were actually moons—now known as the Galilean satellites in his honor.

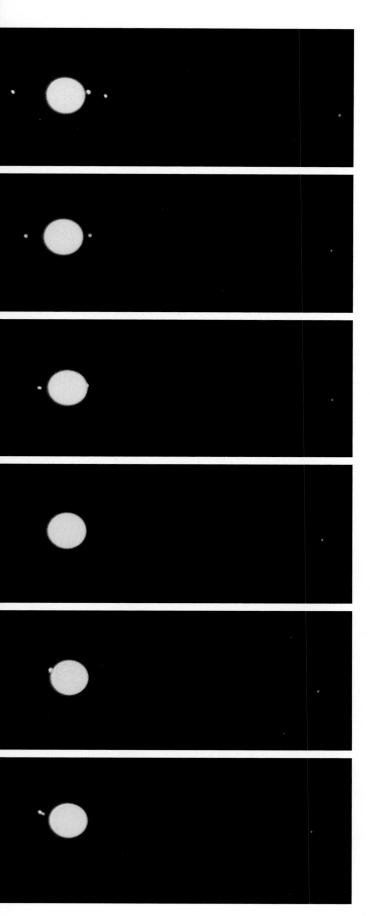

Not all was lost, though. In its dying moments, Phobos the probe transmitted thirty-seven images of Phobos the moon. As revealed by the cameras and the craft's thermal infrared sensors, the satellite's surface did not appear uniformly dark, as had been expected. Rather, it seemed to vary from bluish to reddish in places, suggesting to some analysts that the satellite is not composed entirely of carbonaceous material. Other scientists theorize that the apparent surface variations could be the result of inconsistencies in the instruments themselves or in the way the data were interpreted. *Phobos 2* did make one unequivocal contribution, however. It provided a more precise measurement of the moon's gravity field, lowering the density estimate for Phobos to about 1.95 grams per cubic centimeter. This figure, below that expected even for a C-type asteroid, suggests that Phobos's interior could be porous, which in turn might indicate that the satellite is a jumble of fragments that reaccreted loosely after a shattering impact.

In the end, the Soviet mission raised more questions than it answered. Despite the study that Phobos and Deimos have been subjected to in recent decades, closer examinations with more sophisticated instruments will be required if the experts are to diagnose what the moons are made of and trace their origins.

A SOLAR SYSTEM IN MINIATURE

Unlike the Martian satellites, which long eluded discovery and still defy comparison, the four largest moons of Jupiter were sighted early on in the history of European astronomy, and they were recognized from the start as akin to the planets in the Solar System, with Jupiter playing the part of the Sun. The first to discern this pattern was Galileo Galilei, who trained his telescope on Jupiter and its environs on January 7, 1610, a few months after he first surveyed Earth's moon. Amid the profusion of stars he was seeing for the first time, the astute Galileo at once noted the significance of three small, bright objects around the great planet. They "aroused my curiosity," he wrote in his account, *The Starry Messenger*, "by appearing to lie in an exact straight line" and by being "more splendid" than other stars. He noticed on the next night that all three of them had changed positions and on succeeding nights that they continued to move about Jupiter in varying patterns.

Within a few days, he had solved the puzzle. "I had now decided beyond all question," he wrote, "that there existed in the heavens three stars wandering about Jupiter as do Venus and Mercury about the sun." He soon observed a fourth moon of Jupiter in the same plane as the other three, and rushed his report into print, naming the group the Medici planets in honor of his patron, Cosimo II de' Medici. The announcement proved no less controversial than Galileo's assertion that Earth's moon was far from the perfect body it was reputed to be. Here was proof that heavenly objects revolved around a focal point other than Earth, which in turn supported the claim of Galileo and others that Earth was just one of several planets circling the Sun—an idea still vehemently opposed by the Church. One cautious traditionalist even refused to look at the satellites through Galileo's telescope on grounds that the instrument was bewitched.

A different kind of challenge arose in Germany from an astronomer named Simon Marius, who claimed to have sighted the moons first. Whatever the truth of this assertion, Galileo was first to document the discovery, and he remained so closely identified with the four satellites that they came to be known as the Galileans. Yet Marius left a legacy of his own. The moons acquired the individual names he assigned them—those of four figures seduced or abducted by the Roman god Jupiter in classical mythology. Marius dubbed the outermost satellite Callisto, and the other three in order toward the planet, Ganymede, Europa, and Io.

Although the brightness of the satellites suggested that they might be bodies of considerable size, neither Galileo's telescope nor the more powerful

GALILEAN GEOLOGY

The surfaces of Jupiter's four Galilean moons suggest that the satellites have quite different histories. Callisto and Ganymede are pocked with ancient impact craters, a sign that no internal geologic activity has resurfaced these outer moons in eons. But Europa and Io are virtually crater-free. Io's nearness to massive Jupiter subjects it to strong tidal pulls that cause molten material to bubble up from the interior, covering old scars; on Europa, the molten matter may primarily be heated by radioactive decay in the core.

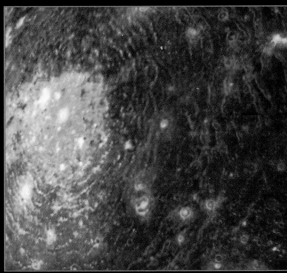

Callisto. The dirty, ice-covered face of Callisto *(above)* is completely saturated with impact craters, showing no sign of internal geologic processes. White splashes *(right)* are areas of fresh ice exposed by relatively recent impacts.

lenses of his successors revealed anything of their surfaces. The moons remained objects of scientific interest, however. In the late 1600s, Danish astronomer Ole Römer used Io to arrive at a calculation of the utmost importance. Römer noticed that the amount of time between the occultations of Io—the moon's periodic disappearances behind Jupiter—increased as Earth's orbit took it farther from Jupiter. Rejecting the prevailing assumption that the speed of light was infinite, Römer correctly attributed the fluctuating intervals between occultations to the varying distances that reflected light from Io had to travel before reaching the observer. By timing the intervals through the year, Römer was able to gauge how long it took light to cross the diameter of Earth's orbit—the first measurement of the speed of light and one that turned out to be accurate within about 30 percent.

AND THEN THERE WERE FIVE

Over the next two centuries, observations of the four distant satellites accomplished little more than to define their orbits more precisely, confirming that the moons follow nearly circular paths around their primary while inclining only marginally to its equatorial plane. This orderly Galilean quartet remained undisturbed until 1892, when American astronomer E. E. Barnard, a diligent observer after the manner of Asaph Hall, used the thirty-six-inch refractor at the Lick Observatory in California to spot a fifth moon quite near Jupiter. Although Barnard's telescope was more powerful than the Naval Observatory refractor Hall had used to spot Phobos and Deimos, his feat of detection was no less remarkable. The new satellite was located at a distance from Jupiter equivalent to just one and a half times the planet's

Ganymede. Varying crater density indicates Ganymede's past geologic activity. Dark, heavily cratered regions contrast with lighter areas where impacts have largely been resurfaced. As on Callisto, white splashes are newly exposed ice.

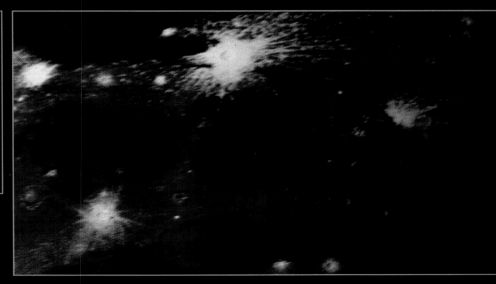

radius and was all but lost in its glare, presenting an even fainter target than the Martian pair. Barnard's find was later dubbed Amalthea, after the mythical goat who suckled Jupiter as a baby. Although the moon's orbit was similar to those of the four Galileans in shape and inclination, Barnard's discovery turned out to be merely the first of several finds over the next few decades that would show Jupiter's satellite system to be far less regular in structure than the Solar System it supposedly resembled.

The moons discovered after Amalthea were detected through the analysis of photographic plates, a technique perfected by asteroid hunters. The observer would attach a camera to the telescope and expose the film for long periods of time, then search the developed frame with a special viewing instrument for an anomalous streak of light standing out against the predictable pattern of the background stars. All the Jovian satellites discovered in this manner lay beyond Callisto, at the outer edge of the planet's gravity field, and in traditional fashion, all received the names of the mythical god's lovers. The first two of Jupiter's new consorts were sighted by Charles Perrine, an Ohio businessman and photographer who joined the Lick Observatory staff as secretary in 1893. He found the planet's sixth and seventh moons—Himalia in 1904 and Elara in 1905. Three years later, at the Greenwich Observatory, British astronomer P. J. Mellote discovered the eighth satellite, Pasiphae. Then a tenacious tracker from Illinois, Seth Nicholson, claimed a streak of finds, beginning with Sinope, which he happened to see on a photograph he took while trying to view Pasiphae at the Lick Observatory in 1914. Nicholson later moved on to the thirty-six-inch refractor at the Mount Wilson Observatory in California, where he sighted Lysithea and Carme in 1938, and

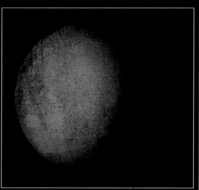

Europa. Intricate fractures crisscross the smooth, ice-covered landscape of Europa, suggesting surface renewal that may be driven by heat from radioactive decay and stresses resulting from Jupiter's gravitational pull.

Ananke in 1951. All these moons are quite small and trace highly elliptical orbits around Jupiter; those with names ending in "e" lie farthest out and move in retrograde fashion (opposite Jupiter's spin direction), while those with names ending in "a" are grouped somewhat closer to the planet and follow prograde paths.

AN ORGANIZING PRINCIPLE

By 1951, when Nicholson's discovery of Ananke brought Jupiter's roster of satellites to twelve, it had become abundantly clear that the moons of the Solar System as a whole were a more diverse lot than the planets. Yet astronomers lacked a method for classifying the bewildering profusion of satellites—which now numbered thirty-one, including nine around Saturn, five around Uranus, and two around Neptune. Into the vacuum stepped a bold theorist named Gerard Kuiper, a Netherlands-born astronomer and one of the pioneers of planetary science in the United States. He argued that the jumble of moons could be best organized by sorting them into two major categories: the regular and the irregular.

The regular group was typified by the Galilean satellites, whose circular, equatorial, prograde orbits suggested that they had formed around Jupiter in much the same way that the planets had around the Sun. According to theory, the Solar System emerged from a vast nebula of gas and dust that was gradually compressed by gravity into a disk of material swirling about a central point. There at the hub—the area of greatest density—the proto-Sun took shape. Elsewhere within the disk, matter condensed and combined over the eons to yield planets, orbiting in the same plane and in the same direction

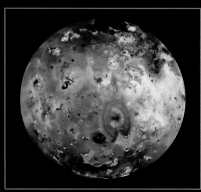

Io. Dotted with volcanoes fueled by tidal stresses, Io is the most geologically active moon known. The volcano Loki *(right)* spews material from the left end of its elongated fissure, just above a horseshoe-shaped pool of liquid sulfur.

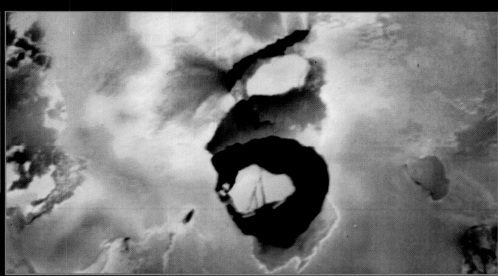

as the original disk. A regular satellite system such as Jupiter's, Kuiper argued, originated in a similar manner from a disk of gas and dust swirling about the emerging gas giant. Evidently, the Sun's gravity and heat prevented this process from operating to similar effect around the inner planets. Although Kuiper defined the two Martian moons as regular based on their orbits, scientists later questioned whether any of the relatively small, rocky inner planets ever possessed the sort of extended gaseous nebula required to foster regular satellites. As for Earth's moon, Kuiper conceded that it was an anomaly—fairly regular in orbit, but too large relative to the planet to have been produced by the same mechanism that spawned the Galilean moons and others of their class.

The irregular satellites were a different breed entirely. In this outlaw group, Kuiper placed moons with one or more deviant traits: highly elliptical orbits, significantly inclined orbits, or retrograde paths. (Some of Jupiter's outer moons exhibit all three characteristics.) Kuiper believed that many such irregulars were ordinary moons that in a sense had gone wrong. He theorized that as satellites formed within a swirling nebula, their growth served to reduce the mass of the protoplanet, which in turn allowed moons at the system's periphery to escape their parent's gravitational embrace and begin orbiting the Sun. Such stray satellites would follow a path similar to that of the planet, setting the stage for a future encounter that would reintroduce the moon to its primary but at an odd angle, producing an eccentric, inclined, or retrograde orbit. Astronomers later dispensed with all but the last part of this scenario, arguing that most of the irregular moons were simply asteroids or comets that had approached their respective planets at various inclinations and speeds and ended up in sundry orbits.

In the process of being captured, the intruders sometimes underwent punishing changes, as evidenced by the two groups of irregulars around Jupiter *(pages 84-85)*. The outermost clan of retrogrades—Ananke, Carme, Pasiphae, and Sinope—are so tightly bunched and follow such similar paths that they may well be the remnants of a single intruder that was fractured by a collision. Another such breakup may explain the evident similarities between the prograde irregulars. For decades, astronomers assumed that there were just three members of that clan: Himalia, Lysithea, and Elara. But then in 1974, astronomer Charles Kowal, using a combination camera and reflector at California's Palomar Observatory, made out a fourth such consort, Leda. Smaller than Phobos, this dark fragment remains one of the faintest objects in the Solar System.

WEIGHING THE REGULARS

Kuiper's ranking system was tidy, and his ideas about the origins of moons seemed to fit the facts astronomers were amassing about the big Galilean satellites of Jupiter. Well before the sophisticated Voyager probes reached those moons in 1979 and relayed images of their surfaces, ground-based observers came up with data supporting Kuiper's analogy between the evo-

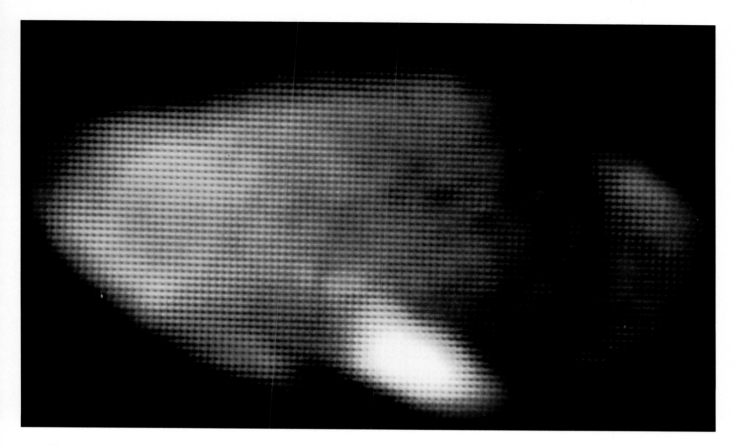

Amalthea. Looking like a ruddy potato in this color-enhanced *Voyager 1* photo, Jupiter's Amalthea is covered with sulfur—ejected by neighboring Io—and darkened by interaction with Jupiter's magnetosphere. The moon is irregular in shape and, like the two satellites of Mars, orbits with its longest axis pointed toward its planet. The brighter region may be underlying material newly exposed on the slope of a large impact crater.

lution of the Galileans and the development of the planets. In the early 1970s, astronomers obtained the first accurate measurements of the diameters of the Galilean moons by tracking them as they occulted background stars of known magnitude and distance. As was suspected, the four satellites were quite large; Ganymede, the biggest of the lot, turned out to be wider than the planet Mercury, and Europa, the smallest, was nearly the size of Earth's moon. By comparing the volume of each satellite to its mass—derived by tracking variations in the orbits of the moons as they perturbed one another—researchers were able to gauge their density with unprecedented precision.

Density was the key value, because it offered a rough idea of whether a moon was predominantly rocky or made up of lighter stuff, such as liquids or gas. Kuiper had argued that the inner Galilean moons, Io and Europa, ought to resemble the innermost planets in being quite dense and rocky. The reason for this, Kuiper explained, was that Jupiter—which even now radiates twice as much energy as it receives from the Sun—was considerably hotter in the turbulent early stages of its formation, and its energy would have gradually evaporated water and other volatiles on the nearby satellites, leaving them hard and parched like Venus and Mercury. By contrast, the outer moons, Ganymede and Callisto, would have retained more of their volatiles and should be less dense.

The new figures seemed to bear out this projection. Io, the closest in, was indeed the densest of the four Galileans, weighing in at a solid 3.5 grams per cubic centimeter. Europa, next out, appeared slightly less rocky at 3 grams per cubic centimeter. And the values for Ganymede and Callisto were lower still, at about 1.9 grams per cubic centimeter each. Pondering the pattern,

In the Throes of Gravitational Tides

As moons revolve around their planets, they are squeezed and stretched by the same force that spawns ocean tides on Earth: the differential gravitational pull exerted on the nearer and farther sides of one celestial body as it orbits another. Because moons are small compared with their planets, they suffer much larger tides than do their planets. The surface of Earth's moon, for example, bulges three times higher in response to the planet's pull than does Earth's surface in response to the Moon's tug. On Io, the innermost of Jupiter's four Galilean satellites, powerful periodic tides raise the surface as much as 200 feet.

Frictional heat generated by this stress causes the moon's interior to heat up, much as a tennis ball grows warm if it is repeatedly flexed. Generally speaking, the degree of tidal flexing and the resulting heat depend on the size and shape of the moon's orbit. The smaller its orbit—that is, the closer the moon is to its planet—the stronger the gravitational forces acting on it and the higher its tides. Similarly, the more eccentric a moon's orbit, the greater the difference in its distance from the planet at its closest and farthest points (pericenter and apocenter, respectively) and the more pronounced the difference between the amplitude of the tide at these points. In addition, the difference in the moon's orbital speed at pericenter and at apocenter causes the tidal bulges to shift back and forth across its surface, warming the moon even more.

Tidal heating, along with the heat generated by phenomena such as radioactive decay in a moon's core, can produce dramatic transformations in the body's interior and on its surface, as shown here in a hypothetical planet-moon system. Each of the planet's three moons orbits outside the Roche limit and is shown at four different positions along a color-coded path representing the moon's movement over the course of thousands or millions of years of tidal interaction with the planet.

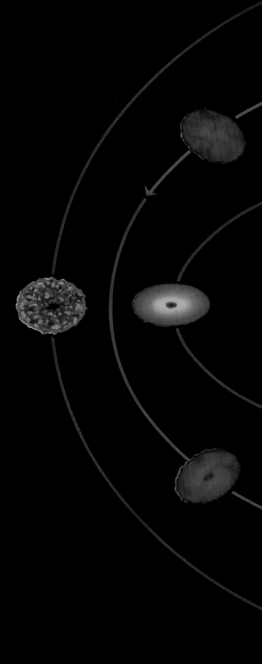

The innermost moon in the system shown here *(green path)* is caught in a cosmic tug of war between its parent planet and nearby sibling moons, causing its orbit to become more eccentric. The eccentricity increases the variation in the height of the moon's tidal bulges at different points in its orbit, generating heat that eventually leads to rifting and faulting of the crust and the eruption of volcanic material. The process is most dramatic on Jupiter's Io, where volcanoes dwarfing any on Earth spew plumes of debris that can reach more than 200 miles into space and flows of molten material spread over vast areas, covering the craters that once pocked its surface.

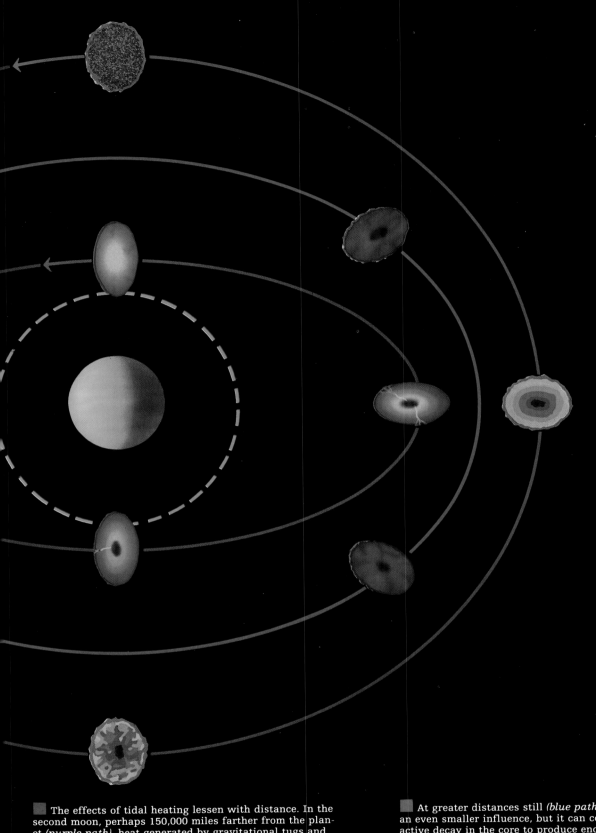

The effects of tidal heating lessen with distance. In the second moon, perhaps 150,000 miles farther from the planet *(purple path)*, heat generated by gravitational tugs and radioactive decay is not sufficient to produce volcanic activity but may prevent the moon's surface from hardening completely. The resulting crust is thin and somewhat fluid, and may be periodically broken by oozing flows of magma. The process may explain why Jupiter's second moon, Europa, which is covered with a thin crust of cracked ice that should retain the marks of impacting meteorites, appears to be nearly crater-free: Water seeping from beneath the ice fills and covers the craters.

At greater distances still *(blue path)*, tidal heating has an even smaller influence, but it can combine with radioactive decay in the core to produce enough internal heat to keep a moon somewhat fluid. Over thousands of years, such fluidity allows heavier materials to sink toward the center of the satellite and lighter components to migrate outward to form the crust. On bodies such as Ganymede, Jupiter's third Galilean moon, which is thought to have been composed originally of a rocky water-ice mix, this differentiation might have caused slushy ice from the mantle to rise to the surface, forming the ridges that cover much of the moon's frozen face.

astronomer John Lewis of the Massachusetts Institute of Technology took Kuiper's ruminations a step further. He concluded that Ganymede and Callisto had the right density for bodies composed of rock and water in roughly equal measures; the water should exist predominantly in the form of ice, he ventured, but some melting could be expected due to the radioactive decay of core elements and other factors. Presumably, Europa would be a drier body than the two outer moons, and Io would be a wasteland not unlike Earth's moon. Yet Lewis and others noticed puzzling signs that not all was quiet on Io. In 1973, Robert A. Brown, a research fellow at Harvard's Center for Planetary Physics, conducted a routine spectroscopic analysis of the light reflected by Io simply to confirm that his instruments were in working order before he analyzed the spectra of the other Galilean satellites. To his surprise, Brown detected the signature of sodium in Io's spectrum. Later observations revealed the source—a fine cloud of sodium atoms accompanying the moon in its orbit around Jupiter—and other astronomers also spotted sulfur in Io's path. Clearly, something strange was going on there.

The activity on Io was still unexplained when the twin Voyager probes were boosted into space from Cape Canaveral, Florida, in the late summer of 1977. Then in early March 1979, just days before *Voyager 1* was due to encounter Jupiter, three researchers published a daring article in the journal *Science* that suggested a solution to the Io enigma. Stanton Peale of the University of California, along with Patrick Cassen and Ray Reynolds of NASA's Ames Research Center, proposed that Io had a hot, molten core, fueling volcanoes that ejected gas into the atmosphere and constantly resurfaced the moon with magma. This would give the satellite a fresh and uncratered crust, very different from Earth's Moon. Io was being heated, they suggested, by the combined gravitational effects of Jupiter and the other Galilean moons. The outer satellites perturbed Io's orbit significantly, and the resulting eccentricity compounded the tidal disruptions caused by massive Jupiter, which tugged at Io with greater or lesser force, depending on the distance between the two bodies. As a result, the entire structure of the moon was being flexed back and forth, making it probably "the most intensely heated terrestrial-type body in the solar system," Peale and his colleagues wrote.

DAZZLING NEW VISTAS
On March 5, *Voyager 1* put these radical ideas to the test. As the probe swept around Jupiter that morning at a distance of about

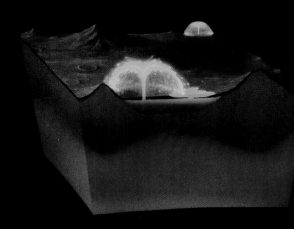

IO'S DUELING VOLCANOES

Almost as surprising as the discovery of active volcanoes on Io was the finding that they are of two distinct types. Some vents spew towering but relatively short-lived plumes of material; the eruptions of others are less spectacular but may last longer.

As illustrated above, molten rock *(red)* under intense pressure in Io's interior often breaches the solid crust *(dark gray)*, where it may heat overlying sulfurous deposits *(tan)* to the boiling point. If the deposit is sulfur dioxide, which vaporizes easily at -10 degrees Celsius, the resulting modest geyser will vent as long as the sulfur dioxide lasts. If the deposit is primarily sulfur, however, the buildup of heat needed to reach its 445-degree-Celsius boiling point may produce a sudden, powerful explosion that subsides relatively quickly.

A false-color *Voyager 1* image of Io captures the volcano Loki in action as it shoots a sulfur dioxide plume about 100 miles into space. The plume—extending from the moon's left edge—was still visible when *Voyager 2* arrived four months later. Io's largest volcano, the sulfur-driven, heart-shaped Pele *(right)*, was active for *Voyager 1*, displaying 200-mile-high plumes, but was dormant during *Voyager 2*'s flyby. Many millions of years of eruptions have blanketed Io with a thick, multicolored sulfurous residue.

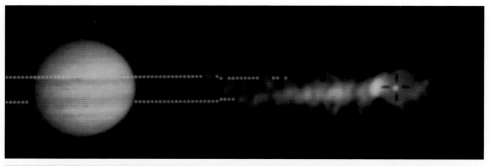

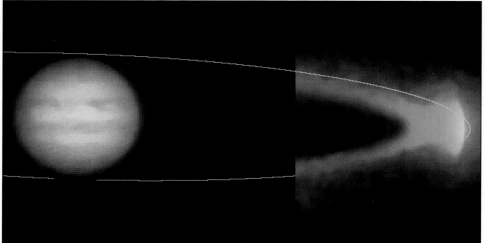

Io's volcanic venting and the interaction of the moon with high-energy particles in Jupiter's magnetosphere combine to generate distinctive clouds, as shown in the two photographs at left; in both, the image of Jupiter has been superimposed for scale. In the top photo, electronically neutral material—mostly sodium atoms—has settled into a vast cloud orbiting Io. Cross hairs mark Io's position within the cloud, and the yellow line traces the shared orbit of the moon-cloud system around Jupiter. In the bottom image, electronically charged material—primarily sulfur ions—forms a torus around Jupiter, held in place by the planet's magnetic field. This false-color edge-on view reveals a cool inner region *(green)* and a hot outer region *(purple)*. The densest part of the torus is marked in yellow.

a half-million miles and raced on toward Saturn, it passed within 15,000 miles of Io and relayed astonishing closeups. Scientists and reporters who had gathered at NASA's Jet Propulsion Laboratory (JPL) in Pasadena to review the incoming images groped for words to describe the moon's eerie landscape, aglow with hot shades of yellow and orange. Some characterized the sight as "grotesque," but Bradford Smith, leader of NASA's Imaging Team, remarked charitably that "Io looks better than a lot of pizzas I've seen." The moon's lurid surface was covered with dark spots that the scientists were hard-pressed to interpret; one prominent blemish lay at the center of a heart-shaped formation more than 600 miles wide. These spots did not look like conventional impact craters—indeed, there were no obvious craters anywhere on Io—but the experts were reluctant to identify the dark areas as volcanoes without solid evidence to back up that exotic possibility.

Voyager 1 caught only distant glimpses of the next Galilean moon, Europa. It looked bright, icy, and remarkably smooth, with the exception of faint streaks across its surface. For more details, the scientists would have to await the closer approach of *Voyager 2.* As it was, *Voyager 1* gave the instruments at JPL all they could handle. On the evening of March 5, the probe flew by Ganymede at a distance of about 70,000 miles, returning haunting views of a mottled and frozen terrain marked with dark and light patches of ice, slender fault lines, and crisp craters ringed with bright streaks of ejecta. The next morning, *Voyager 1* passed nearly as close to Callisto, whose icy surface appeared darker and dirtier than Ganymede's and was absolutely peppered with craters. One enormous impact basin stood out. Its central crater, nearly 1,000 miles across, was surrounded by concentric ridges. Scientists dubbed the area the "bull's-eye."

In addition to providing breathtaking views of Jupiter's major satellites, *Voyager 1* treated astronomers to a few bonuses. On the day before its encounter with Io, it got a close look at the tiny inner moon Amalthea. Although regular in orbit, the moon proved to have far less in common with the four Galileans than with the peculiar Martian pair, Phobos and Deimos. Dark with a reddish tint and shaped like a lumpy potato that had been pared down by impacts, Amalthea measured less than 100 miles wide and about 160 miles long. As far as Amalthea was concerned, *Voyager 1* showed astronomers more or less what they had expected to see, but the probe sprang a surprise that same day when an eleven-minute exposure taken by one of its cameras afforded a glimpse of a thin band of ring particles circling the planet's equator. The find, announced on March 7, established Jupiter as the third planet after Saturn and Uranus known to be endowed with a ring system.

No less dramatic was the discovery made on March 8 by a young optical navigation engineer at JPL, Linda Morabito. Poring over images on the electronic screen, she noticed a crescent-shaped cloud on Io's horizon. As she and her colleagues worked over the data, it became clear that they had located the plume of a huge volcano. The more they looked, the more plumes they found. By the end of the search, they had pinned down eight active hot spots, among them the heart-shaped formation with a dark area in the center that proved to be Io's largest volcano and earned the title Pele. Here was confirmation of the theory put forward by Peale and his coauthors—that Io had a hot, molten core and was continuously spewing magma and gases from the dark vents that scarred its surface. As *Voyager 1* sped away toward Saturn, project scientist Ed Stone remarked without overstatement that the probe's brief encounters with Jupiter and its satellites had yielded "almost a decade's worth of discovery."

Like a cleanup batter, *Voyager 2* arrived four months later to finish the job. It appeared that Jupiter's environment had changed significantly. Sensors detected a lower concentration of sulfur and other particles in the area, hinting that Io's volcanoes had quieted down. In fact, most of the active vents spotted by *Voyager 1* were still spouting; one of the volcanoes, dubbed Loki, had even thrown up a second plume. But the giant Pele, which in March had been spewing a fountain of ejecta more than 200 miles high, now lay dormant.

Voyager 2 offered its most exciting revelations when it passed less than 130,000 miles from Europa on July 9. Now the moon's innumerable streaks stood out in crisp relief, revealed to be ridges and cracks lacing the otherwise smooth surface. Assessing the geologic damage to the satellite, Laurence Soderblom, deputy leader of the Imaging Team, remarked that it was as if "Europa had been cracked, broken, by some process which crushed it like an eggshell and just left the pieces sitting there. Expansion and contraction of ice and water are a good way to crunch up the surface." Apparently, there was enough heat in Europa's interior to keep its icy shell relatively thin and brittle. *Voyager 2* also detected signs of surface cracking and shifting on Ganymede, but elsewhere the moon's frozen crust was thick enough to have absorbed powerful meteorite impacts without fracturing. Finally, the probe confirmed that outermost Callisto was the coldest and least active of the four Galilean moons, as evidenced by its heavily cratered surface, which must have been dead for billions of years to retain so many impressions. "There's just not room for another crater on that body," Voyager analyst Garry Hunt marveled. "It's totally full."

A PATTERN UNFOLDS

In the wake of the Voyager encounters, scientists intent on tracing the evolution of Jupiter's major moons had to return to the drawing boards. Plainly, Kuiper and the theorists who followed in his path had identified only a few of the variables in the cosmic equation that produced the Galileans and other regular satellite systems. As NASA's Bradford Smith put it, the Voyager data "shattered the dogma that the worlds of the outer solar system had been shaped by similar, predictable, and rather uninteresting processes." Yet the new picture was not so complex as to resist interpretation. With the help of the Voyager insights, scientists came up with modified views of the early history of the Galileans, seeking to explain both their basic similarities and their striking differences.

One recent model of the evolutionary process begins with the assumption that the four satellites accreted from a mix of rocky materials, including iron and radioactive elements such as uranium, and water ice—abundant around Jupiter and the planets beyond thanks to their distance from the Sun. As the satellites grew, the gravitational forces exerted by each sphere increased, with tidal effects on Jupiter that may have caused the planet's own gases to heat up. Jupiter's radiant energy left nearby Io with little water, while the other moons retained more depending on their distance from the planet. As the Jovian system stabilized, the planet cooled, and Europa, Ganymede, and Callisto assumed their permanent frosty exteriors.

If the evolutionary process had halted there, the Galileans would not have displayed nearly so many idiosyncrasies as the Voyager probes detected. But other forces conspired to agitate the moons and warm their interiors to varying degrees. The heat produced through tidal flexing *(pages 72-73)* had a pronounced effect on Io, as evidenced by its ongoing volcanism, and may also have helped crack the surface of Europa, which likely retains a liquid ocean beneath its icy crust. Ganymede, significantly larger than the inner moons and farther from Jupiter, probably experienced little in the way of tidal heating, and Callisto even less. But the big outer moons were not without alternative sources of thermal energy—notably the decay of core radioactive elements—and the faint warmth from the interior along with frequent impacts endowed them with some distinctive surface features.

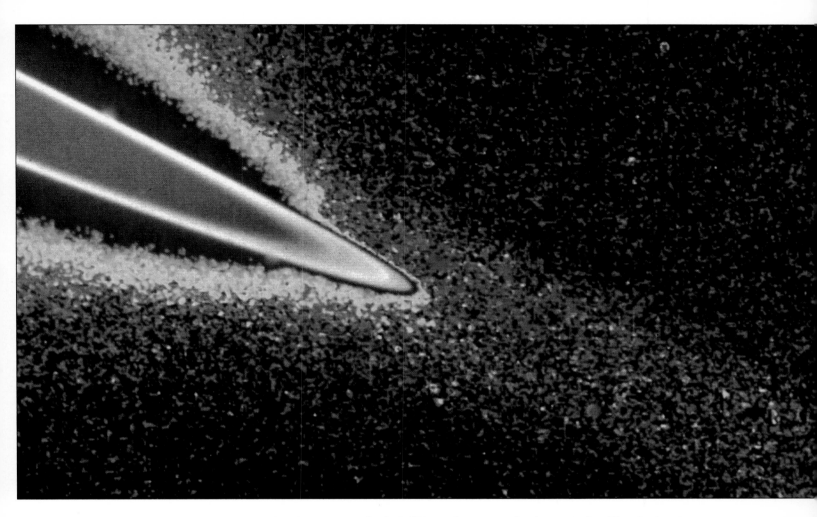

Rings around Jupiter. A false-color image from *Voyager 2* allows scientists to discern the three components of Jupiter's evanescent ring system. Inside the bright, sharply defined main ring *(white)*, a cloud of particles known as the halo *(red)* flares to embrace the planet. (Jupiter is out of the image, to the left.) Outboard of the main ring, the very faint gossamer ring *(blue-green)* traces its barely detectable arc.

Much of crater-pocked Callisto, for example, is coated with what astronomers have described loosely as "dirty ice," apparently the accumulated debris of countless incoming meteorites large and small. For all the battering Callisto has endured, it possesses few craters wider than 100 miles and none that have steep ridges at the perimeter like the craters on Earth's moon. This may indicate that Callisto once possessed enough internal heat to allow its surface ice to flow, thus filling in the deeper impressions. Ganymede presents a different puzzle for planetary scientists in its curious patchwork surface, alternating "dirty" regions like those on Callisto with areas that appear quite clean. The largest continuous dark zone, located on the side of the moon that always faces away from Jupiter, is about the size of the continental United States. Named the Galileo Regio, it has the densely cratered look of Callisto, a sign that the region has long been inactive. The light areas, on the other hand, are coated with what looks like fresh ice and other signs of surface activity, including long grooves that are separated by ridges. The grooves are several miles wide and the ridges rise roughly 3,500 feet—approximately the height of the Appalachian Mountains. Apparently, heat emanating from Ganymede's core produced these folds in the icy crust along with actual cracks in the surface. When the fissures opened, water surged up from below and froze at the chill surface temperatures, forming a shiny new coating of ice. The apparent absence of such resurfacing on Callisto suggests that Ganymede was endowed with a warmer core, containing more radioactive elements.

Europa's shiny surface with its fractured look testifies to an even hotter core there, likely kindled by a combination of radioactive decay and tidal flexing. If, as scientists suspect, Europa's mantle is warm enough to preserve an ocean of water beneath its icy shell, then the tepid depths could conceivably harbor primitive life forms such as the algae found in ice-covered lakes in Antarctica. Io, for its part, has a destructive surplus of internal heat. By one estimate, its volcanoes disgorge at least 100 billion tons of material each year, enough to cover the entire moon with a layer of ejecta more than ten yards deep every million years.

Between fiery Io and giant Jupiter lies the planet's wispy ring system *(pages 90-91),* consisting of inner and outer bands of small fragments, divided by a gap, and a diffuse halo of even finer particles. As researchers scrutinized the Voyager photographs after the Jovian encounters, they discovered three additional tiny moons tucked in among the rings—named Adrastea, Metis, and Thebe—which brought the total number of satellites orbiting Jupiter to sixteen. Mere points of light on the images, these dwarfs were estimated to be less than one-third the size of Amalthea. Scientists concluded that they were fragments of larger objects that had been smashed by flying debris or perhaps pulled apart by tidal forces. Their location strongly suggested that they play some part in the orchestration of Jupiter's rings. But a clearer understanding of the relationship between such moons and the particles they accompany would not come until later, when the Voyager probes reached the brilliant ring world of Saturn and opened up another episode in their kaleidoscopic survey of the outer Solar System.

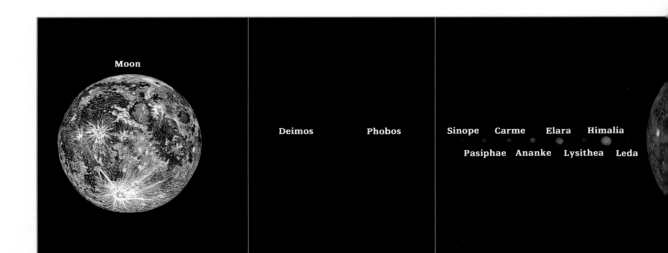

SATELLITE CENSUS

Astronomers have long thought of the main asteroid belt (broad white band above) as a natural boundary, separating the small, rocky planets such as Mars *(red)* from huge Jupiter *(orange)* and the other gaseous worlds beyond. The contrast between the inner and outer regions of the Solar System applies not only to the makeup of the planets but to the abundance of their moons, as revealed by a census of the satellite families on either side of the main belt: Mars possesses just two companions, while Jupiter's collection numbers sixteen *(below)*.

This discrepancy reflects differences in the Sun's influence on planetary formation. The Sun's intense heat left the four inner planets compact and dense, and its gravity apparently robbed them of the extended nebulae of gas and dust needed to foster regular satellite systems. A collision likely equipped Earth with its one unusually large moon, while Mars acquired its strikingly small and craggy pair under mysterious circumstances *(page 82)*. If Mercury and Venus ever had satellites, powerful solar tides must have caused those moons to crash into their parent planet or break free from it altogether.

By contrast, Jupiter and its big outer neighbors Saturn, Uranus, and Neptune were far enough from the Sun to amass large followings of satellites. Jupiter's gas-rich primordial embryo spawned a quartet of regular moons—sizable bodies with nearly circular, equatorial orbits—ranging from innermost Io to outermost Callisto. Farther out lie eight small irregular satellites: captured bodies with eccentric orbits highly inclined to the planet's equatorial plane. Inside Io lurk four moonlets classified as collisional shards, leftovers from ancient smash-ups that orbit within the planet's delicate ring system.

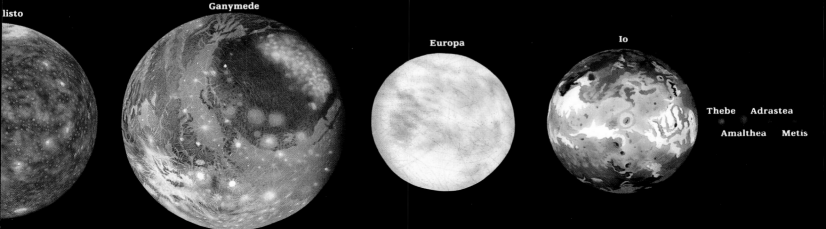

Callisto   Ganymede   Europa   Io   Thebe   Adrastea   Amalthea   Metis

Battered Phobos is some seventeen miles
across at its widest. The gaping crater Stickney and the long grooves emanating from it
were carved by an impact that nearly shattered the moon.

A Peculiar Tandem

For almost a century after the discovery of the faint Martian moons Phobos and Deimos in 1877, astronomers could do little more than define their orbits. The tiny bodies traced circular, equatorial paths around Mars and could thus be classified with the regular moons of the outer planets. But in recent decades, as earthbound instruments and robotic probes taught scientists more about the trajectories, shapes, and surfaces of the satellites, the Martian moons began to look increasingly peculiar—an odd couple with few parallels, if any, in the Solar System.

For one thing, they have opposite orbital destinies. As diagramed at right, Deimos orbits at a distance 6.9 times the radius of Mars from the planet's center *(outer blue line)*, putting it beyond the synchronous orbit distance *(dashed line)*, the point at which a moon would complete one orbit in the time it would take Mars to spin once on its axis. Deimos thus finishes its circuit in thirty hours and eighteen minutes, or nearly six hours longer than the Martian day. The tidal interactions resulting from this lag are very gently pushing Deimos outward *(pages 62-63)*. For Phobos, whose orbit lies at 2.8 planetary radii—well inside the synchronous orbit distance for Mars—just the opposite is true. Circling the planet in just seven hours and thirty-nine minutes, the inner moon is being drawn inexorably toward Mars and is bound ultimately for destruction.

The two moons are unusual in shape and substance as well. Scarred by countless impacts, they look more like crude building blocks leftover from the Solar System's infancy than like fully formed satellites. Indeed, their dark, disfigured appearance, combined with their low density, has led some astronomers to argue that they began their careers as asteroids and were later captured by Mars when they deviated from the main asteroid belt, perhaps kicked out by gravitational interactions with massive Jupiter.

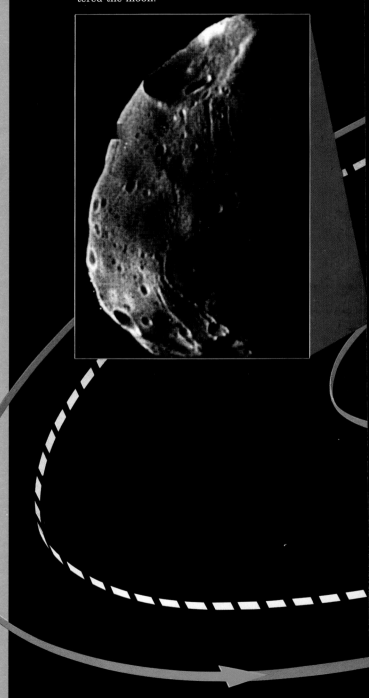

The smoother surface of nine-mile-wide Deimos indicates that it possesses more surface debris than its inner neighbor—and that it has not recently suffered an impact strong enough to dislodge that dusty coating.

THE REVEALING ORBITS OF A SPRAWLING CLAN

Jupiter's extended family includes all the satellite types known to astronomers. This impressive brood owes its size and diversity to the planet's favorable position in the early days of the Solar System—far enough from the Sun's searing radiation to retain a dense envelope of gas and dust. That endowment not only supported the formation of Jupiter's four regular moons but lent the planet sufficient mass to capture satellites at great distances and to sustain a ring system replete with several moonlets.

Jupiter's regular moons—born of a disk-shaped nebula circling the protoplanet—inherited rounded, equatorial, prograde orbits *(below, right)*. Although the four inner moonlets trace similar paths *(left)*, two

lie so close to the planet that they could not have accreted there. Wherever they originated, the moonlets all appear to be vestiges of larger bodies, smashed by cosmic projectiles such as comets or asteroids and left to sweep about Jupiter in tight circles like ancient hulks caught in a whirlpool. Astronomers believe that these shards endure pummeling from smaller objects in their vicinity, providing a fresh supply of dusty debris for the planet's tenuous ring system.

The orbits of Jupiter's eight distant irregular moons *(right)* tell a different story. Since their paths are highly eccentric and inclined to the planet's equator, they could not have emerged from Jupiter's primordial nebula; they must have formed elsewhere and then been captured by the planet's gravity. As if by design, they fall into two groups of four—one prograde *(green)* and the other retrograde *(orange)*. Each quartet may have a common ancestor: a larger moon that broke apart after being captured.

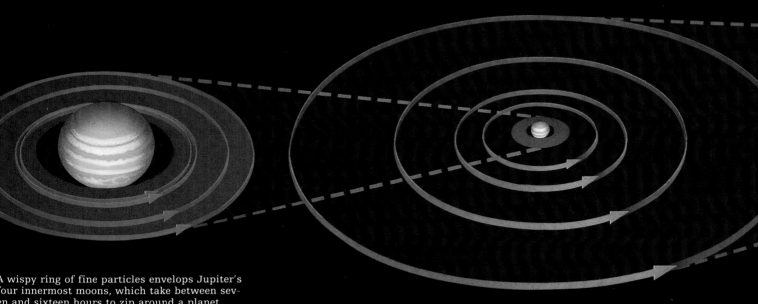

A wispy ring of fine particles envelops Jupiter's four innermost moons, which take between seven and sixteen hours to zip around a planet that completes a single rotation in just under ten hours. At the ring's outer edge is Thebe, 3.1 Jovian radii from the planet's center. Next, at 2.5 radii, is the quartet's largest member, Amalthea, followed by two tiny, closely spaced satellites at about 1.8 radii, Adrastea and innermost Metis. The ring itself extends inward to around 1.4 planetary radii—just 18,000 miles from Jupiter's cloud tops.

Farthest of the four regular moons, Callisto orbits at twenty-six planetary radii from Jupiter's center and takes 16.7 days to circle the planet. Ganymede, next in, lies fifteen radii from the planet. Closer to Jupiter are two smaller satellites: Europa, at nine radii, and Io, at just six, or nearly as close to the giant planet in absolute terms as the Moon is to Earth. The orbital periods of Ganymede (7.2 days), Europa (3.6), and Io (1.8) have evolved in such a way that all three moons can never be on the same side of Jupiter at once.

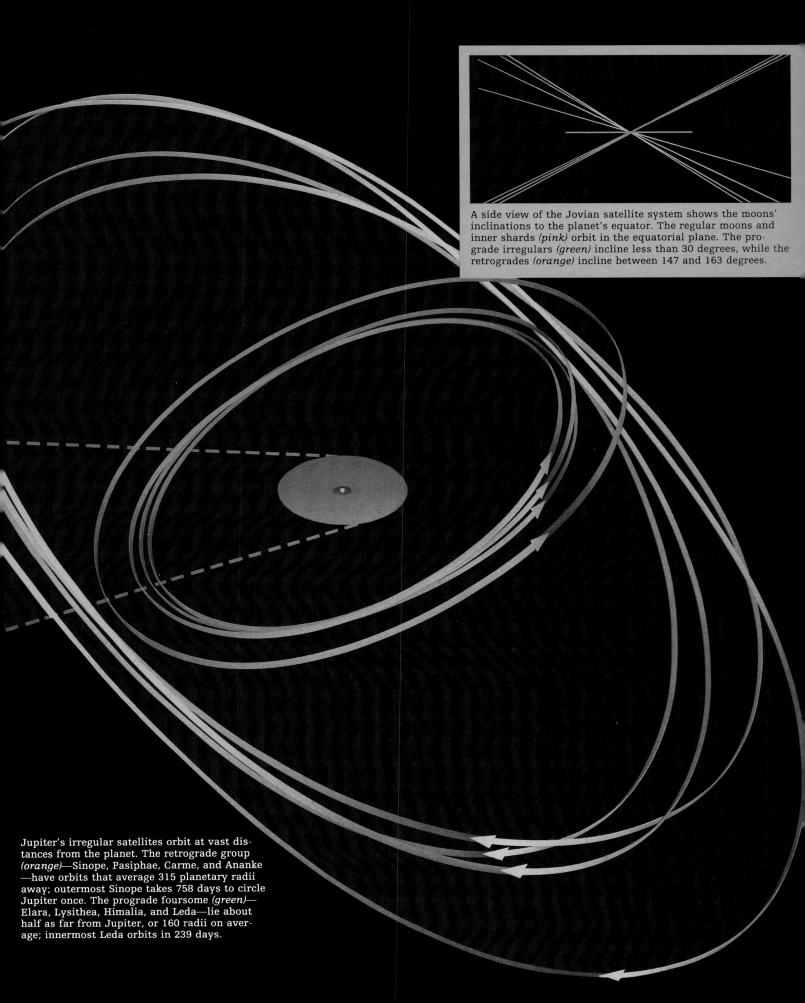

A side view of the Jovian satellite system shows the moons' inclinations to the planet's equator. The regular moons and inner shards *(pink)* orbit in the equatorial plane. The prograde irregulars *(green)* incline less than 30 degrees, while the retrogrades *(orange)* incline between 147 and 163 degrees.

Jupiter's irregular satellites orbit at vast distances from the planet. The retrograde group *(orange)*—Sinope, Pasiphae, Carme, and Ananke —have orbits that average 315 planetary radii away; outermost Sinope takes 758 days to circle Jupiter once. The prograde foursome *(green)*— Elara, Lysithea, Himalia, and Leda—lie about half as far from Jupiter, or 160 radii on average; innermost Leda orbits in 239 days.

THE ICY ALLOTMENTS OF CALLISTO AND GANYMEDE

In recent years, astronomers have identified some intriguing parallels between the planets of the Solar System and the four regular moons of Jupiter, known as Galileans for their discoverer. As with the planets, the makeup of the Galileans varies significantly with their distance from the primary. The two farthest moons, Callisto and Ganymede *(bright pink orbits above)*, are significantly less dense than Europa and Io, and although the contrast is not as sharp as between the gaseous outer planets and the rocky inner ones, the difference stems from the same source: heat. In this case, radiation from a proto-Jupiter of starlike intensity apparently prevented water and other volatile substances from condensing, leaving the inner moons rockier and denser than those farther away.

Spared the worst of this grilling, Callisto and Ganymede emerged with roughly as much ice as rock. Originally, those two basic constituents were mixed together, and they might have remained so had it not been for two additional sources of heat: repeated violent impacts and decaying radioactive elements in the rock. As a result of these influences, Callisto and Ganymede began to differentiate: The heavier rocky material descended to form a core while the melted ice surfaced to form deep layers of water that eventually refroze partially if not completely.

Callisto. With a diameter of 2,980 miles, Callisto is significantly larger than Earth's moon and even more thoroughly cratered, having endured countless impacts that darkened its ice with meteoritic debris. In fact, Callisto's surface is so densely pitted that it must have remained undisturbed by internal processes since shortly after the moon took shape, some four billion years ago. Nonetheless, there is evidence that beneath its ancient icy crust *(white, far right)*, Callisto harbored a partially fluid mantle *(blue)* and a rocky core *(brown)* capable of generating some heat. For one thing, Callisto's biggest impact sites, such as Valhalla *(near right)*, are marked by systems of concentric ridges that look very much like ripples frozen into the surface of a pond—an indication that the crust was undergirded by something more yielding than pure ice. In addition, the moon's craters are surprisingly flat, suggesting that enough warmth emanated from the interior to allow the ice to flow at a glacier's pace and even out depressions. Although Callisto is colder today than in the past, it may still possess a slushy layer beneath its icy crust.

Ganymede. At 3,270 miles across, Ganymede is slightly larger and a bit denser than Callisto, which may help explain why Ganymede has undergone more heat-driven geologic upheaval. About two-fifths of Ganymede's surface *(near right)* resembles Callisto's in that it is dark, heavily cratered, and very ancient. But the remaining terrain is grooved and light-colored, and has fewer craters. Early tectonic processes triggered by warmth from the rocky core *(far right, brown)* might have cracked Ganymede's frozen crust *(white)* and sent water from the slushy mantle *(blue)* gushing up through the fissures to resurface the area with fresh ice. Whatever cooling Ganymede has experienced since these episodes, it still possesses a thinner crust than Callisto and may retain a relatively thick layer of slush.

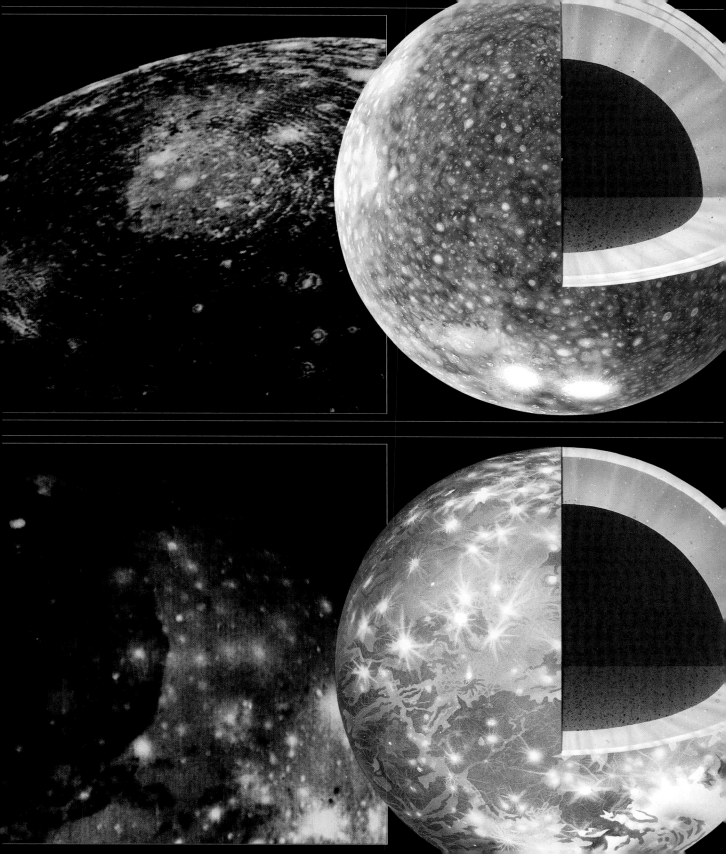

INNER MOONS WITH WARMER UPBRINGINGS

Compared to Callisto and Ganymede, which retained a good deal of ice and grew to considerable size, Europa and Io *(bright pink orbits above)* took shape in a hotter environment close to Jupiter and ended up proportionately smaller and rockier. Both are only about two-thirds the size of Callisto or Ganymede, but Europa is one and a half times as dense and Io nearly twice as dense as either of the outer moons. Io's proximity to Jupiter rendered it a virtual desert; what little water the moon might have contained originally was baked out and lost to space long ago.

Although Jupiter's radiation diminished as the planet matured, Europa and Io continued to be heated by the decay of radioactive elements in their cores and by tidal friction resulting from the gravitational tug of the planet and the perturbing effect of the outer moons. Such friction helped endow Europa with a fairly warm interior: Its clean, relatively uncratered crust of ice has evidently been resurfaced within the past 100 million years or so by watery slush pouring through cracks in the surface. The tidal effects have been even more pronounced on close-in Io, whose crust reveals no impact craters at all—a consequence of the fresh material being deposited continuously there by ongoing volcanism.

Europa. Slightly less than 2,000 miles wide, Europa is the smallest Galilean satellite. It is also the brightest, thanks to its relatively fresh icy surface. Yet that glistening crust is deceptive: Europa's high density indicates that its interior is primarily rocky *(far right, light brown)* with some metals perhaps mixed in at the core *(gray)*. Enough heat may emanate from the dense interior to preserve a shallow subterranean ocean *(blue)* below the frozen exterior *(white)*. Scientists speculate that the lines crisscrossing Europa's surface *(near right)* were conduits for the exchange of material between that watery ocean and the crust, with fresh liquid ascending through rifts to resurface the moon while ancient, dirty crustal material descended into the depths through trenches.

Io. Dry, dense, and volatile, 2,200-mile-wide Io is thought to have a solid core *(far right, gray)* surrounded by partially molten silicate rock *(red)* and a thin silicate crust *(orange)* with pockets of sulfur and frozen sulfur dioxide. Subject to four times as much tidal stress as Europa, Io releases its pent-up energy in fierce volcanic outbursts producing towering plumes of vaporized sulfur and sulfur dioxide. The volcano Pele *(near right)*, photographed during the flyby of *Voyager 1* in March 1979, raised a plume 200 miles high *(white wisps)* and deposited fallout *(brown ring)* up to 450 miles away from its central vent *(white fissure)*. Although Io seemed to resemble a pizza in early Voyager images, its true surface colors are more subdued—sulfuric yellows and browns.

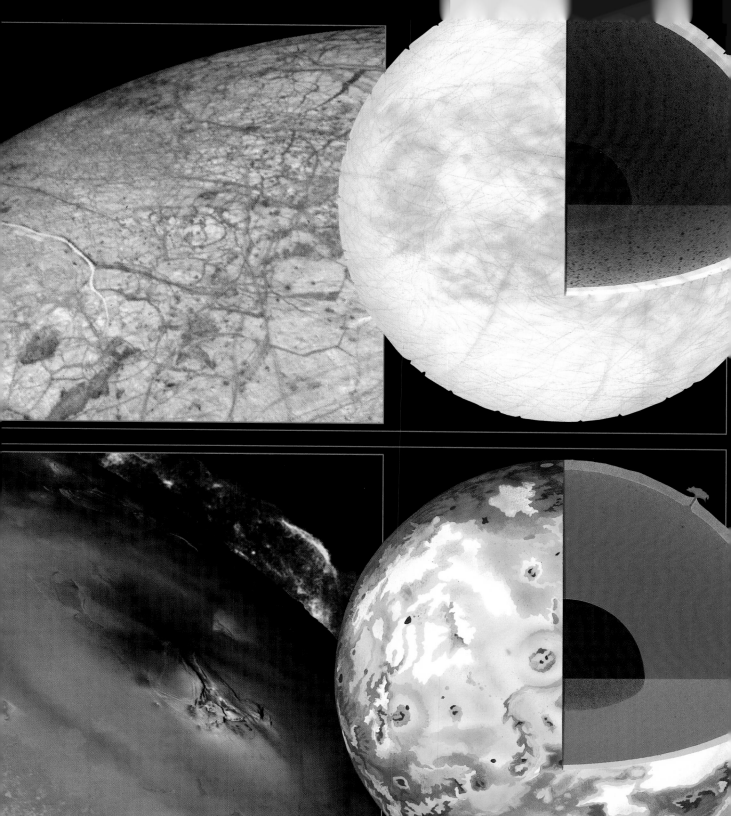

A Ring Replenished by Moonlets

Like the nebula of particles that swarmed about Jupiter in its infancy, most of the dusty specks that make up the planet's present-day ring system trace roughly circular paths about the planet's equator, as do its four embedded moonlets *(above)*. Whether the minute ring fragments are in fact vestiges of the primordial nebula or, as many scientists now believe, more-recent products of the breakup of larger objects lured into Jupiter's gravitational hub, they do not orbit in place eternally. Over time, they are ground into ever-finer particles, measuring less than a millionth of an inch. Yanked out of the equatorial plane by Jupiter's magnetic field, the motes become part of a delicate inner halo that flares out toward the planet's poles.

Because the tiny particles eventually spiral down into Jupiter's atmosphere, some process must be replenishing them. A look at the density pattern of Jovian ring fragments *(right)* offers a clue as to how and where new particles might be entering the cycle. The heaviest concentration—the so-called main band—embraces the innermost moonlet, Metis, and nearby Adrastea, which are about twelve and eight miles wide, respectively. Micrometeorites constantly chip away at their surfaces, and the dusty debris readily escapes their weak gravity, providing fresh material for the main band. By contrast, the bigger outer moonlets, Amalthea and Thebe, have the gravitational capacity to retain such powdery debris, which may help explain why the area beyond the main band, the so-called gossamer ring, is so sparse and tenuous.

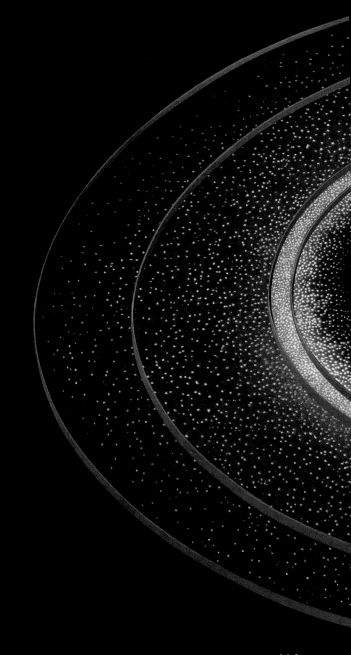

The main band of Jupiter's ring system, which extends from the inner moonlets toward the planet for some 4,200 miles, is only about eighteen miles thick and consists of particles that take between five and seven hours to complete an orbit; they are thus moving faster than Jupiter is spinning and are being drawn slowly inward. Closer to the planet lies the halo, which reaches a thickness of more than 12,000 miles as it flares out. The outer, gossamer ring—a band of negligible thickness—reaches all the way to Thebe, farthest of the four moonlets.

A mosaic of *Voyager 2* images taken in July 1979 from a distance of nearly a million miles captures a backlighted Jupiter and its slender band of ring particles. The microscopic specks that constitute the ring appear quite bright, just as otherwise invisible dust motes gleam when caught in a ray of sunlight.

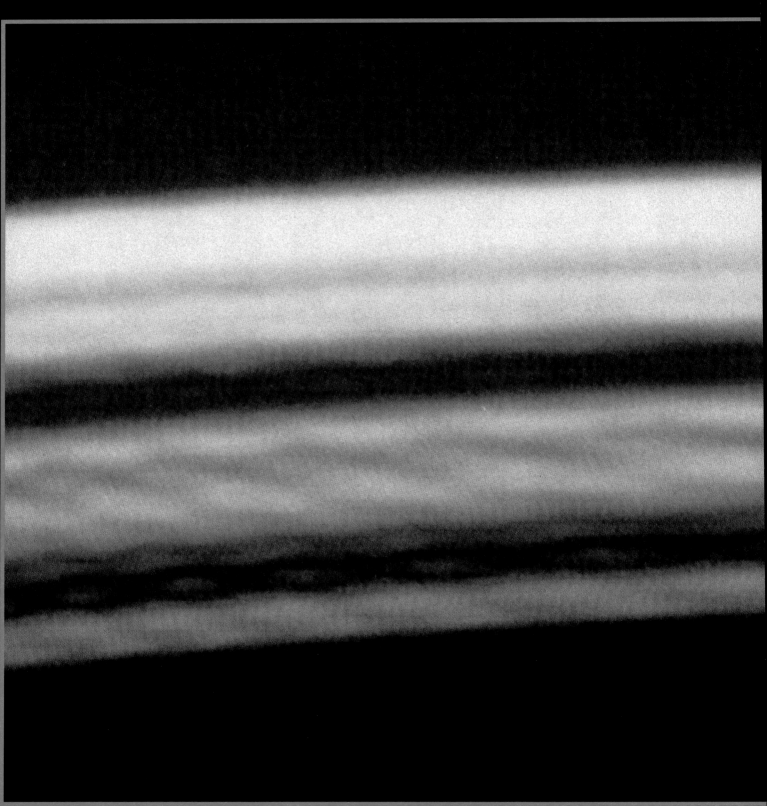

The tenuous filaments of Uranus's Delta ring show up as bright bands separated by dark gaps in a false-color image generated from data returned by *Voyager 2* in 1986. The probe's instruments recorded suc- cessive dips in light coming from a distant star as the rings passed in front of the star. Scientists discov- ered the Uranian rings nine years earlier while monitoring a similar stellar occultation from Earth.

igh above the southern reaches of the Indian Ocean on March 11, 1977, a team of astronomers from MIT pursued a shadow through the predawn darkness. Their chase vehicle was a C-141 military transport that had been converted into an airborne observatory, and the shadow they looked for would be cast by the planet Uranus as it moved in front of a distant star. Data from the so-called stellar occultation promised to be richly revealing. By timing its duration, the astronomers might be able to establish the planet's size, and by analyzing the degree to which the star's light was refracted as it faded, they might gauge the depth—and perhaps the composition—of Uranus's upper atmosphere. But Jim Elliot, leader of the team, was worried; calculations indicated that there was a 17 percent chance the airborne observatory would fail to intercept the track of the occultation. A colleague had joked that in the event of a miss, Elliot should at least claim to have placed an upper limit on the size of a hypothetical Uranian ring, because any such ring would have produced some light blockage. The suggestion became a running gag among the astronomers; everyone knew that Saturn was the only ringed planet.

As special instruments on the plane monitored the positions of Uranus and the star, all seemed to be going well. But thirty-five minutes before the predicted occultation, Ted Dunham, Elliot's graduate student, blurted, "What was that? What was that?" He had noticed a dip in the level of light from the star, as if something had gotten in the way. Elliot immediately suspected high-altitude clouds or a glitch in the star tracker. Yet the instrument seemed to be functioning properly, and there were no clouds. The light level dipped again, then three more times. Referring to the team's joke about Uranian rings, Elliot said quietly, "It looks like maybe we've got one."

Still, the duration of the dips had been so short that it was difficult to believe the minioccultations had been caused by rings; Saturn's, after all, were huge. When the team returned to their base in Perth, Australia, and began studying the unexpected flickers, speculation centered on the notion of a belt of moons around Uranus. But the more the MIT scientists and other astronomers looked at the readings, the clearer the meaning of the flickers became. Elliot and his team had outfitted distant Uranus with rings.

The inadvertent find ushered in a new era in the annals of ring science. Before another decade passed, Jupiter would prove to have rings as well, and

Neptune would have dropped some hints of being similarly accoutered. Plainly, something in the processes that created the gas-giant planets of the outer Solar System also encouraged the creation of rings. Puzzling out the secrets of this ring-trove became a passion for planetary scientists.

At the same time, other astronomical riches came to light. Before the epic journeys of the two Voyager spacecraft, Saturn was thought to possess just nine moons, Uranus five, and Neptune two. These moons all fell into two familiar categories: regular satellites, which evidently formed along with their parent planets; and irregulars, which apparently formed elsewhere and were gravitationally captured by a planet. But some of the satellites spotted by the two Voyagers seemed to belong to a new class of bodies: remnants of shattered moons, scraps and shards of deadly collisions. Although Saturn's Hyperion, for example, had always seemed oddly shaped, astronomers had spent little time pondering the significance of that fact. As the space probes revealed, the history of the outer Solar System was a violent tale indeed. Moreover, even some of the supposedly unremarkable regular satellites were seen to have been transformed by a strange array of forces—ice volcanoes, nitrogen geysers, and exotic atmospheric chemistry.

In a sense, the outer Solar System was only running true to form: It had been demonstrating a capacity to amaze for centuries. One epochal surprise occurred in 1610, when Galileo Galilei first gazed upon the rings of Saturn. With his homemade 30-power telescope, he was unable to resolve details, but it seemed to him that Saturn had a pair of large, prominent ears. He assumed that the ears were actually satellites, such as the ones he had discovered orbiting Jupiter earlier that year. But Galileo's confidence in his explanation must have been shaken two years later when he looked again at Saturn and found that the ears had vanished. "Has Saturn, perhaps, devoured his own children?" he asked. "I do not know what to say in a case so surprising, so unlooked for, and so novel."

Fellow astronomers shared Galileo's bafflement. A variety of explanations for Saturn's curious behavior were put forward, most of them fanciful. The most popular explanation was that a gaseous exhalation from Saturn condensed into an ever-changing cloud that encircled the planet. A few decades later, however, a better theory emerged. In March 1655, the Dutch scientist and optician Christiaan Huygens turned his new, improved telescope to the sixth planet and noted that the mysterious feature first noted by Galileo was tilted about twenty degrees to the plane of Saturn's orbit—roughly the same angle at which the planet's axis tilts. When the object disappeared the following winter, Huygens reasoned that the relative motions of Earth and Saturn combined periodically to present it edge on to a terrestrial observer, rendering it invisible. Huygens concluded that Saturn was encircled by a "thin, flat ring, nowhere touching, and inclined to the ecliptic."

He believed that the ring was solid, but other scientists doubted that such a thin structure could survive the stresses produced by its rotation around Saturn. Italian-born astronomer Giovanni Cassini, then the director of the observatory of the Académie Royale des Sciences in Paris, believed that the only explanation for the ring was that it consisted of a "swarm of small satellites," each particle orbiting the planet exactly the way a tiny moon would. Cassini strengthened the case for a "corpuscular" ring with his discovery in 1675 of a dark line that seemed to divide the ring into two separate sections. A solid ring would be fatally weakened by such a division. Many more gaps were later glimpsed.

As late as the mid-nineteenth century, some astronomers continued to believe that the rings were solid. Others suggested that they might be fluid, that is, gaseous or liquid. But in 1848, the French theorist Édouard Albert Roche dealt a serious blow to both ideas. He calculated that, within a certain distance from the center of a planet—approximately 2.5 planetary radii—a fluid satellite would not only be torn apart by tidal action but would not have been able to form in the first place. (No fluid satellites have ever been found, of course.) Solid satellites that formed outside this distance, which became known as the Roche limit, might conceivably survive inside the boundary, provided they were strong enough structurally to withstand the tidal pull and were not too large. But they could not form there, because tidal action would prevent the potential constituents from accreting. Since Saturn's visible rings all lie inside the Roche limit and presumably had formed there, they could not be either fluid or solid.

A few years later, the Scottish scientist James Clerk Maxwell used his prodigious mathematical powers to make the case for Cassini's alternative. "The only system of rings which can exist," he concluded, "is one composed of an indefinite number of unconnected particles revolving around the planet with different velocities according to their respective distances."

Definitive proof that the rings were particulate was provided in 1895 by James Keeler, the thirty-eight-year-old director of the Allegheny Observatory near Pittsburgh. That spring, Keeler was measuring the rotational velocity of the planets through the use of spectroscopy. The principle of his investigations was simple: One edge (or "limb") of a rotating planet would be moving toward an observer, and the other would be moving away. These velocities would show up as equal but opposite Doppler shifts in the spectrum of light reflected from the planet, with spectral lines of the receding limb shifted toward the red end of the spectrum and lines from the approaching limb shifted toward the blue. The rotational velocity of the rings of Saturn could be determined in the same way. But that was not all. On April 9, as he was making a two-hour spectroscopic exposure of Saturn, Keeler suddenly realized that his readings could confirm the particulate theory. If the rings were composed of a myriad of small bodies, each with its

Cassini Division. In 1676, Giovanni Cassini published this drawing of a dark gap he discovered in what was thought to be a single ring around Saturn. The division, which now bears Cassini's name and separates the A ring from the B ring, was the first evidence of fine structure within Saturn's ring system.

Keeler's proof. A page from James Keeler's 1895 letter to colleague George Ellery Hale describes the experiment by which Keeler confirmed what he called the "meteoric constitution," or particulate nature, of Saturn's rings. Keeler showed that the spectrum of light reflected by the rings varies in a way appropriate to individual objects orbiting at different distances from the planet and hence at different velocities.

own orbital velocity, then the degree of Doppler shift would vary continuously across the rings, from the swift inner particles to the slower outer bodies. As soon as his spectrogram was developed, Keeler saw that this was indeed the case. The ring reading, he told a colleague, is "the prettiest application of Doppler's principle that I have ever seen."

Astronomers were now certain of the makeup of Saturn's rings; observational evidence and theoretical considerations could all be satisfied by a swarm of small particles independently orbiting Saturn in a thin equatorial plane. The particulate nature of the rings even seemed to explain their gapped structure. Ring scientists theorized that the gaps resulted from a certain type of gravitational interaction between ring particles and Saturn's moons. Particles at specific radial distances from Saturn would orbit in a period that is a simple fraction of the orbital period of a distant moon—two orbits for a particle versus one for the moon, three for a particle and two for the moon, and so on. As a result of this so-called resonance, any given particle at these distances would be subjected to the moon's gravitational attraction at fixed points in its circuit. Over long stretches of time, the rhythmically repeated gravitational tug of the moon would pull the particle into a different orbit, effectively evacuating the path it once occupied *(pages 106-107).* When applied to uncountable individual particles, scientists believed, the process created the observed gaps.

A more basic issue was the origin of the particles, but astronomers also had a ready answer for that question. The little chunks of matter were leftovers from the era of planetary formation—relic material that had somehow avoided being incorporated into either Saturn or its moons.

MORE RINGS

Saturn's rings now seemed satisfactorily explained, and scientists thought little more about them in the decades that followed, except to wonder why only Saturn was so endowed. Although the discovery of the rings of Uranus put that perplexity to rest, it also raised a new set of questions.

Probably as the result of a titanic collision with another body, Uranus is tilted much more severely than Saturn or Earth, inclining 82 degrees to the ecliptic. Thus, at various points along its orbital track around the Sun, its rotational poles are aimed almost directly at Earth. Such was the case in 1977 when the rings were spied. In all, nine were found, concentrically arrayed around the planet's upended equator, like the circles around a bull's-eye. This orientation made them much easier to detect during an occultation than if they had been presented edge on. Still, it had been a lucky sighting. While Saturn flaunts broad rings, Uranus wears only the stingiest stuff—and astronomers wanted to know why.

In 1977, Saturn was thought to possess three main ring systems: the inner C ring, thin and diffuse, extending from 8,700 miles to 19,900 miles above the planet; the dazzling B ring, reaching outward 35,400 miles from the planet; and, beyond the 2,175-mile-wide Cassini Division, the 9,300-mile-wide A ring.

(Other reported sightings, of an interior D ring and a diffuse E ring, were matters of controversy.) In stark contrast, the rings of Uranus are so dark and narrow as to be virtually invisible. The nearest to the planet, ring 6, is about 9,300 miles above the Uranian clouds; the most distant, the Epsilon ring, is situated about 15,500 miles from the planet. Six of the rings are slightly inclined to the planet's equator and have orbits that are slightly eccentric, or out-of-round. The most eccentric ring is also by far the most massive, the outermost Epsilon. It varies in radial width from about 12 miles to 56 miles, while the others have widths ranging from about .6 mile to 7.5 miles.

Scientists barely had time to ponder the peculiarities of the Uranian rings before *Voyager 1* presented them with a third planetary ring system—Jupiter's—to puzzle over. Jupiter's ring had actually been detected during the flyby of *Pioneer 11* in 1974, but nobody was certain of the reading at the time. As the spacecraft flew past the planet, instruments that monitored high-energy charged particles trapped in the planet's magnetic field registered two abrupt drops, one on either side of the planet, as if something was blocking the charged-particle flow. The incidents were generally attributed to an unknown electromagnetic effect, but investigators Mario Acuna and Norman Ness of NASA's Goddard Space Flight Center suggested an alternative: "Although we consider it remote, the possibility exists that the two minima are due to sweeping effects by an unknown satellite or ring of particles."

In 1977, following the discovery of the Uranian rings, Tobias Owen and Candice Hansen of the Voyager Imaging Team persuaded their colleagues to allot time for a single eleven-minute ring-search exposure during *Voyager 1*'s 1979 flyby of Jupiter. Team leader Bradford Smith was skeptical but he was in for a pleasant jolt when the flyby took place. "I had forgotten about it until someone came running in waving the image." The picture had captured what seemed to be a narrow Jovian ring 35,400 miles above the planet's multi-colored clouds. The discovery was so unexpected that it was not announced to the Voyager press corps until the Imaging Team had spent several days verifying that it was not an illusion. Later analysis determined that the ring is some 4,200 miles wide and consists of three main components. The inner-most component, a doughnut-shaped halo of tiny particles, is about 12,000 miles thick; the outermost tapers to a thickness of only 18 miles.

Next, it was Saturn's turn to surprise. The sixth planet unveiled a slew of new ring features for the flybys of *Pioneer 11* (1979), *Voyager 1* (1980), and *Voyager 2* (1981). "Saturn stunned us," Brad Smith later said. The number of major ring components increased to seven, with the confirmation of a wispy interior D ring, and the addition of a weird, narrow, ropelike F ring just beyond the A ring and the diffuse G and E rings extending more than 248,600 miles out from the planet. Moreover, the Voyagers revealed a profusion of ringlets that made the main rings look like a grooved phonograph record. There were ringlets everywhere, even in the supposedly empty gaps. "One gap, the Cassini Division, by itself had more structure than the entire ring system we can see through the telescope," Smith commented. Some of the ringlets were eccentric, some of them were kinky, and the bizarre F ring actually looked as if its various components were twisted around themselves. The Voyagers mapped the ring system down to the scale of a city block, and found complex structure even at that level. Yet for all the intricate architecture, the main Saturnian rings proved to be amazingly thin, no more than a few tens of yards at their thickest point.

With three of the four gas giants now known to possess rings, Neptune became a prime target for ring hunters. In 1984, teams led by André Brahic of the University of Paris and William Hubbard of the University of Arizona used the occultation technique to spot a ringlike feature about 31,100 miles above the planet. But, as viewed from Earth, the ring material filled only about 10 percent of its orbital circumference. Neptune, it seemed, had spurned full encirclement in favor of discontinuous arcs. This would cause much head scratching until *Voyager 2* arrived at Neptune five years later.

RING DYNAMICS
While the explanation for the Neptunian arcs would have to await a flyby, scientists were able to add substantially to their ring knowledge by ongoing detective work from Earth. Using a variety of techniques—including occultations, infrared spectroscopy, and radar surveys—they concluded that the Jovian rings are composed of fine grains of dark silicate dust. This makeup was in contrast to that of Saturn's spectacular rings, which pre-Voyager observations had revealed are mainly water ice, with typical particle sizes ranging from icy dust to snowballs a few inches in diameter to house-size boulders. Now scientists examining the darker rings of Uranus and Neptune suggested that they might be composed of methane ice or some organic material that has been blackened by the effects of radiation. At Uranus, most of the particles are about the size of beachballs, but Neptune's rings have a larger proportion of dust. In all cases, the mass of material making up rings is slight: Even Saturn's spectacular system contains only enough bulk to make a moon 240 miles across, roughly the size of its satellite Mimas.

A mélange of interactions dictate the shape, size, and structure of the ring systems. The classical explanation for Saturn's multiple rings—gravitational resonances between the orbits of ring particles and large outlying moons—turned out to be only part of the story. In another gravitational dance, smaller satellites, moonlets or fragments only a few miles in diameter, are embedded within some rings. By tugging on surrounding particles, these moons can speed up or slow down a given particle, kicking it into a higher orbit or causing it to drop closer to the planet, thereby clearing a path through the ring *(pages 106-107)*. Narrow strands, such as Saturn's strangely warped F ring, are controlled by the gravitational influence of small nearby moons that have come to be known as shepherding satellites. The intricate minuet of large moons, embedded satellites, and shepherds combines with electromagnetic effects and other complex forces to create astonishing architectural variety.

Even as the Solar System's rings have yielded up more of their secrets,

Three images obtained by *Voyager* 2 illustrate the diversity of planetary rings. Saturn *(top)* possesses a bright, broad, complex ring system, in contrast to the rings of Uranus *(center)*, which are relatively dim and exceedingly narrow. The particles making up Neptune's rings are so dark and small that the probe's television camera could discern the bands only by overexposing the brighter orb of the planet, blocked out in the composite image below.

scientists have grown less certain of ring origins and age. Some researchers believe that rings are as old as the planets they encircle, dating back more than four and a half billion years to the birth of the Solar System itself. They were formed, according to this traditional view, from the debris of planetary creation; inside the Roche limit, the ring particles never had the chance to coalesce into moons. But this scenario is flawed for one reason: Rings should not have been able to survive that long.

Over billions of years, gravitational dynamics and collisions with other bodies should have caused the ring particles either to spiral into their parent planet or to spread outward, away from the planet, ultimately dispersing altogether. Arguably, the influence of moons might confine the particles to prevent this dissipation. But if the rings are confined, particle collisions and bombardment by meteoroids should grind the ring components into dust, and dust cannot endure for long before a phenomenon known as plasma drag *(pages 122-123)* sweeps it out of a ring system. Therefore, the dust in any ring system must be fairly youthful. If that is so, however, the system must possess a mechanism for renewing the dust supply. One suggestion is that while collisions might be responsible for rendering large particles small, some collisions may result in the reaccretion of smaller particles into larger ones. These would continue the cycle: rocks to dust and back again. A variation on this theme is that more dramatic collisions—moons hitting other moons or being struck by meteoroids—if they occurred within a planet's Roche limit, would produce debris that could not reaggregate into a moon. Instead, it would spread out into a ring. Saturn's dazzling ring system may well be the product of a moon smash-up that took place within the past billion years.

MANY MOONS

With at least eighteen satellites, Saturn is the most moon-rich planet in the Solar System. The latest was discovered in the summer of 1990 by scrutiny of dozens of *Voyager 2* images received during the flyby nine years earlier. That encounter had revealed a wavy ripple, like the wake of a speedboat, along the edges of the so-called Encke Division in Saturn's A ring. In the mid-1980s, Jeffrey Cuzzi and Jeffrey Scargle of NASA's Ames Research Center had suggested that the ripple indicated the presence of a shepherding satellite or an embedded moonlet and had predicted its orbital characteristics. Finding it, however, would be like locating a needle in a haystack. In 1990, Mark Showalter of NASA's Ames Research Center devised a computer program to pore over the Voyager pictures. The computer found the needle. Measuring less than 12.4 miles in diameter, the moon is the smallest known Saturnian satellite. It was given the temporary designation 1981S13, pending approval of the name Pan, after the Greek god of shepherds.

The little moon is one of at least eleven Saturnian satellites that do not fall into the regular category. The major, regular satellites of Saturn include (proceeding outward from the planet) Mimas, Enceladus, Tethys, Dione, Rhea, Titan, and Iapetus. Beyond Iapetus, the small satellite Phoebe follows a retrograde and highly inclined orbit, characteristics that, added to its dark gray color and low reflectivity, suggest that Phoebe is a captured comet or asteroid. Unlike the major moons of Jupiter, the moons of Saturn are not neatly arranged according to composition and density. Rather, the Saturnian moons are arrayed in a random jumble of sizes and densities, suggesting that satellite formation at Saturn was far from orderly. A higher rate of in-falling comets in the outer Solar System may have been a prominent factor in shaping the satellite system of Saturn—and those of Uranus and Neptune, as well.

If they achieved nothing else, the Voyager encounters left no doubt about the world-hammering violence that prevailed in the early history of Saturn. Ganymede, Callisto, and Earth's own much-battered satellite may resemble targets in a shooting gallery, but Saturn's moons look more like survivors of saturation bombing. In most cases, their bright, icy surfaces are scarred by shoulder-to-shoulder impact craters. Although each moon seems to have a degree of individuality, such as the long trench nearly encircling Tethys, these spheres of ice and rock lack the internal heat necessary to drive an active surface geology. The most distinctive feature on any of them is an immense impact crater on Mimas, with a diameter nearly one-third that of the moon itself. Formally labeled Herschel, the feature became known during the *Voyager 1* encounter as the Deathstar crater, after its uncanny resemblance to Darth Vader's battle station in the film *Star Wars.* The impact that formed the crater must have come very close to shattering Mimas completely, producing more ring material and remnant moonlets.

Iapetus, 900 miles in diameter, is the outermost of Saturn's regular satellites and has long puzzled earthbound astronomers, beginning with Giovanni Cassini, who could not understand why its leading hemisphere is dark while its trailing hemisphere is bright. Some astronomers surmise that Iapetus is sweeping up dark material, perhaps shed from neighboring Phoebe, as it moves along its orbital path. Yet the dark regions on Iapetus are redder than the surface of Phoebe, and Voyager images revealed a 62-mile-wide ring of dark material straddling the border between the dark and bright hemispheres, suggesting that some internal process may have been responsible for spreading a dark stain across half of the moon's icy surface. One recent—and highly disputed—proposal suggests that sometime in the last 100 million years, a comet may have struck the leading hemisphere of Iapetus, dissipating ices while depositing dark, tarlike organic material. For now, however, scientists have reached no firm conclusions about this unique, two-faced body.

Surely the strangest Saturnian satellite is Enceladus, just 310 miles in diameter and only slightly denser than water ice. Such a small, cold body ought to be geologically dead, its frozen surface preserving ancient craters intact and unaltered. But *Voyager 2* stunned scientists with images of what seemed to be an active, youthful surface, gleaming with a new coat of ice. Very few impact craters were in evidence—none at all in some areas—and at least six different, geologically distinct types of terrain were identified. In places, Enceladus resembles Europa, Ganymede, and Mars. There are smooth, un-

Accretion theory. According to this scenario, a primordial cloud of gas and dust *(below)* condensed into all parts of the nascent planetary system simultaneously, in the span of about 1,000 years. Inside the planet's Roche limit, material that was prevented from accreting by strong tidal forces remained in orbit around the equator in a thin, flat disk and is essentially preserved today as a planetary ring.

Breakup theory. In this scheme, the planetary nebula *(above)* coalesced into a center world and a system of moons over the course of perhaps 100,000 years *(near right)*. Later, a meteoroid rocketed through and collided with a satellite inside the planet's Roche limit, breaking the moon apart. Unable to reconsolidate, the orbiting fragments wore each other down, forming a doughnut-shaped hoop *(far right)*. Over the course of millions of years, this torus would flatten and spread into a thin disk *(overleaf)*.

THE ORIGINS OF PLANETARY RINGS

Astronomers currently invoke two rival hypotheses to explain the evolution of the flat, concentric bands of rock and ice particles that gird all of the large, gaseous planets in the outer Solar System. According to the accretion theory *(left)*, a planet, its moons, and its rings all formed simultaneously and relatively quickly from a planetary nebula. As most of the central concentration of gas and dust gravitationally collapsed to form the planet's body, material at the cloud's periphery collected into satellites. Near the surface of the new world, inside the boundary called the Roche limit, tidal forces prevented some of the primordial matter from accreting into larger bodies. Instead, it spread into a flat disk rotating around the planet's equator—a precursor to the elaborate bands that would make up the planet's ring collection.

The alternative view, called the breakup theory *(bottom)*, holds that rings are the pulverized remnants of satellites. In this scenario, the planetary system developed hundreds of times more slowly, with moons forming throughout the nebula. Some then migrated inside the Roche limit through a variety of dynamic processes. Later, one (or more) of these close-in satellites was shattered, most likely by a passing meteoroid. At the mercy of tidal forces, the moon fragments could not reaccrete but rather ground each other down to chunks the size of mountains within a few years, dispersing into a torus orbiting in the former satellite's path. The following pages describe the mechanisms believed to have transformed the torus into a smooth, flat disk of small particles, and then to have carved the gaps that created the individual rings.

Within a few years of the breakup of a close-in moon, fragments had collided with each other and fractured many times over, orbiting the planet in a torus inside the planet's Roche limit.

The chunks steadily bumped together and wore each other into smaller and smaller pieces. Periodically, stray meteoroids crashed into the fragments, further reducing their size, even as dynamic processes *(boxes)* caused the torus to begin to flatten out.

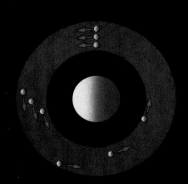

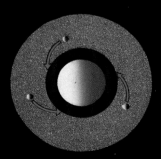

Viscous spreading. As adjacent disk particles travel at different orbital speeds, collisions result in losses of energy and an outward transfer of angular momentum, causing the particles to disperse and flattening the disk.

In-falling matter drag. Material from beyond the planetary nebula collides with particles in the torus, making new fragments and increasing the particle density *(left)*. Because the material rains in uniformly from all directions,

there is no increase in angular momentum, and the greater density causes the particles to lose energy and altitude, a process called orbital decay. The dispersal of disk particles flattens the torus further *(right)*.

FROM TORUS TO RING DISK

In the breakup theory of ring formation, the mountain-size fragments of shattered moon orbiting in a torus inside a planet's Roche zone collided and fractured until they were whittled down to smaller and smaller sizes. As they jostled together, eroding like pebbles gently grinding one another to silt in a stream, the rocks were occasionally smashed by meteoroids and were eventually worn down into particles ranging from several yards to fractions of an inch in diameter.

Collisions among orbiting particles not only reduced the size of the fragments but also caused them to disperse, flattening the torus into a disk *(below)*. Other dynamic processes aided the transformation from torus to disk. One, known as viscous spreading, results from the different speeds at which discrete bodies orbit a planet. According to the rules of orbital mechanics, a body closer to its primary travels faster than one farther away. As particles orbiting near one another collide, the collisions impart angular momentum to particles in higher orbits and sap the momentum of those in lower orbits. The changes in energy, especially in vertical motion, cause the particles to disperse over time *(box, far left)*, making for a flatter torus.

Another process that flattened the proto-ring involved gas and debris in the outer reaches of what was left of the planetary nebula. Orbiting the planet more slowly than the torus material, this matter exerted drag on the orbiting particles, sapping their energy of motion and forcing them to drop into closer orbits *(box, near left)*.

After several hundred or a few thousand years, the narrow torus that defined the orbit of a forgotten moon was transformed into a disk thinner and flatter than a phonograph record *(below)*.

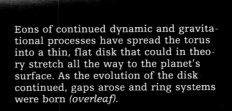

Eons of continued dynamic and gravitational processes have spread the torus into a thin, flat disk that could in theory stretch all the way to the planet's surface. As the evolution of the disk continued, gaps arose and ring systems were born *(overleaf)*.

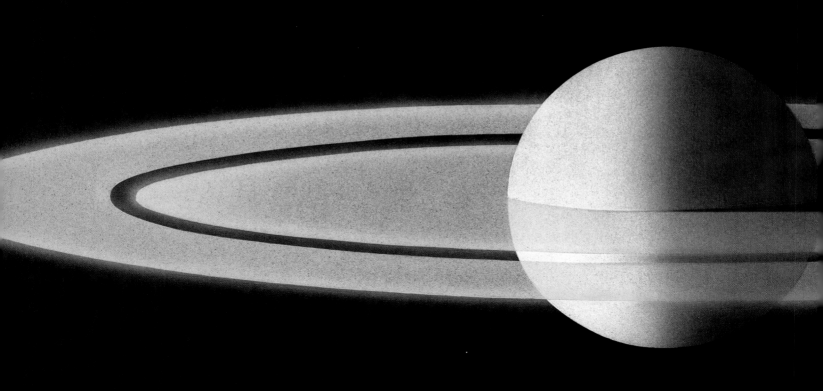

CLEARING GAPS WITH EMBEDDED MOONS

Although the major gaps between rings may be largely the product of gravitational resonances with moons lying outside of the ring system, many of the smaller, more intricate divisions are believed to have been cut by moonlets or large fragments embedded among the particles in the rings themselves.

As an orbiting moonlet overtakes particles farther out from the planet and is overtaken by those closer in, it exerts a strong local gravitational effect, pulling nearby particles slightly out of their orbits *(box, right)*. As a result of the encounter, an individual ring particle either gains or loses energy and shifts into a higher or lower orbit, a process that effectively clears away particles that were once orbiting near the moon-

let's path—producing a narrow gap in the disk.

Because embedded moonlets are so tiny, this scenario was purely hypothetical until recently. In 1985, scientists combing through images transmitted by *Voyager 2* when it flew by Saturn in 1981 found wavy-edged rings that seemed to be indirect evidence for the existence of moonlets; the waves were thought to have been generated by the moonlets rather in the way a motorboat generates a wake. Then, in 1990, astronomers reexamined the images and actually found the moon itself in Saturn's 200-mile-wide Encke Division. At 12 miles in diameter, the satellite—Saturn's eighteenth—had managed to carve a trail more than sixteen times its own width.

By slowing or boosting the orbital speed of particles traveling near it in the ring, a moonlet acts as a sort of plow, clearing a swath through the particles orbiting a planet in a flat disk, without touching any of them.

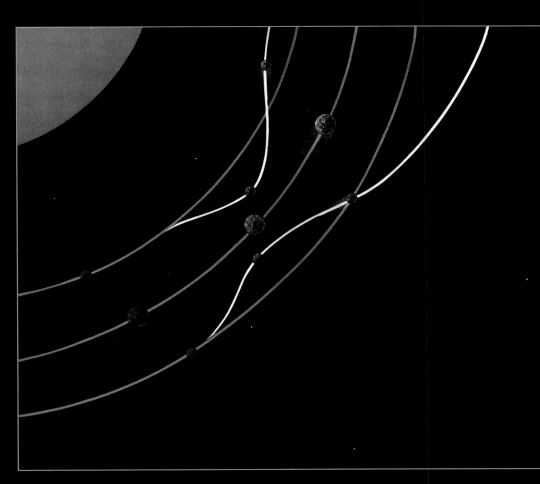

Creating a gap. The gravitational dynamics of orbiting bodies allow a moonlet to sweep its orbit clean of smaller ring particles. As a fast-moving particle on the planet side of the moonlet catches up with and passes the slower-moving satellite, the moonlet's gravity exerts a backward tug that causes the particle to lose momentum. This in turn weakens its resistance to the even greater pull of the planet, and the particle drops to a lower orbit.

Conversely, as a moonlet catches up with and passes a slower-moving outer particle, the moon's gravity pulls the particle along with it. This energy boost flings the particle into a higher orbit.

cratered plains, crumpled ridges, and flow patterns suggestive of an ongoing process of resurfacing by fresh ice from the interior. The most highly reflective body in the entire Solar System, Enceladus may harbor active volcanoes that spew an icy mix of water and ammonia. Scientists puzzled over how such a tiny, frozen world could generate the energy needed to fashion this bizarre landscape. One theory suggests that because Enceladus is orbitally coupled with Dione, making one circuit for every two by Dione, the frequent lineups between the two moons might force Enceladus into a slightly eccentric orbit that could create tidal friction capable of heating the satellite and powering the ice volcanoes.

Titan, discovered by Christiaan Huygens in 1655, was long thought to be the largest moon in the Solar System—until Voyager dethroned it. Although the moon is a bit larger than the planet Mercury, Voyager revealed it to be slightly smaller than the Jovian moon Ganymede. The inability of earthbound astronomers to precisely gauge Titan's size was understandable: It is the only moon with a dense atmosphere. Titan's impenetrable clouds have hidden its surface details even from the prying eyes of the Voyagers, and the moon's geology and topography will remain unknown until the Cassini mission, a probe scheduled for launch in 1996, surveys it with radar sometime early in the twenty-first century.

Previous observers had speculated that Titan might possess an atmosphere since it appeared massive and cool enough (judging by its distance from the Sun) to retain a dense gaseous envelope. But the first to find proof of an atmosphere there was the Dutch-American scientist Gerard Kuiper. He discovered the telltale signature of methane in Titan's spectrum in 1944. For the next three and a half decades, planetary astronomers debated issues such as the atmospheric density, the possible presence of gases other than methane, and the temperature at the satellite's surface. None of these questions could be answered with assurance from Earth-based observations. Finding the answers was considered so important that Titan was designated as one of the highest-priority science objectives for *Voyager 1*'s Saturn flyby in 1980. In order to get the best possible look at Titan, mission planners sacrificed an opportunity to send *Voyager 1* on to Uranus. As it turned out, no one regretted the decision.

On the eve of the first Voyager encounter, there were two competing models of the Titanian atmosphere. One of the models, championed by Robert Danielson and John Caldwell of Princeton, envisioned an atmosphere that was 90 percent

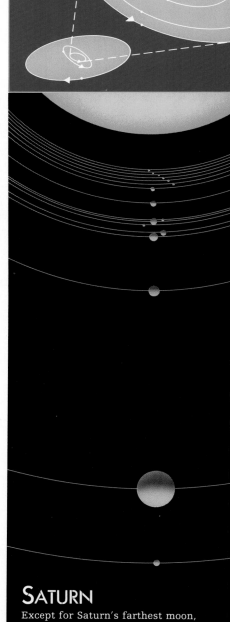

Pan
Atlas
Prometheus
Pandora
Janus
Epimetheus
Mimas
Enceladus
Tethys
Telesto
Calypso
Dione
Helene

Rhea

Titan

Hyperion

SATURN

Except for Saturn's farthest moon, Phoebe, whose path is retrograde and highly inclined, the planet's eighteen known satellites follow fairly regular orbits *(top)*. The bodies themselves, however, are quite diverse, ranging from a swarm of shards near the planet (shown to scale above) to the well-developed world Titan. Some of the moons share orbits, perhaps as a result of an impact that sundered them. The coorbital groups are Janus and Epimetheus; Tethys, Telesto, and Calypso; and Dione and Helene.

Iapetus

Phoebe

methane, with a surface pressure of 20 millibars (Earth's is about 1000 millibars, or one bar) and a surface temperature of just 86 degrees above absolute zero (86 degrees Kelvin). The other model, proposed by Donald Hunten of the University of Arizona, forecast an atmosphere of 90 percent nitrogen, a surface pressure of 20 bars, and a surface temperature as high as 200 degrees Kelvin. The difference was more than a matter of numbers; the two models implied very different sorts of worlds. In Hunten's model, Titan was warm enough to support a huge variety of photochemical reactions—perhaps mimicking the kind of prebiological chemistry believed to have occurred in the atmosphere of early Earth. Titan, some researchers thought, might be the best place in the Solar System to search for life.

Visually, the Titan encounter was disappointing. The ruddy, featureless cloud cover made the world look like what one Voyager scientist described as a "fuzzy, seamless tennis ball." An opaque layer of aerosol particles—probably manufactured by photochemical reactions—shielded the surface from view. But Voyager was versatile, and other instruments revealed what the cameras could not. The spacecraft's ultraviolet and infrared spectrometers confirmed the presence of nitrogen in Titan's atmosphere, as well as methane, carbon monoxide, and a variety of other hydrocarbons. The spacecraft itself, passing just 2,500 miles from the moon's clouds, performed a valuable occultation experiment for scientists on Earth; as it traveled behind Titan, Voyager's transmitted radio signals were gradually attenuated by the atmosphere until they were finally cut off by the surface of the moon. By analyzing the radio data, scientists were able to measure Titan's diameter

Most of Saturn's seven small inner moons serve as custodians of the ring system. Through gravitational interactions, they help maintain the shape of the bands and, by shedding debris when impacted, they may replenish the ring particles. **Pan,** or 1981S13, *(1),* about 12 miles across, orbits in one of the gaps; its existence was inferred from the ripples it generates in the two adjacent rings. **Atlas** *(2),* an oblong body roughly the same size as Pan, was once thought to be a so-called shepherd moon but is now believed too small to exert much gravitational influence on the ring it accompanies. **Prometheus** *(3)* and **Pandora** *(4)* are somewhat larger shepherd moons that work in tandem, corralling their ring from either side of the ring's orbit. The coorbital moons **Janus** *(5)* and **Epimetheus** *(6)* are locked in a gravitational tango, periodically switching positions as they orbit. **Mimas** *(7)*—the largest of the inner moons at 242 miles across—is responsible for clearing part of the Cassini Division, the widest gap in Saturn's rings.

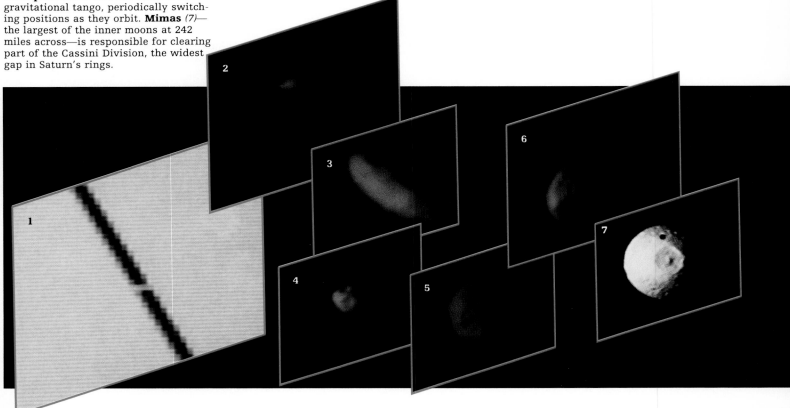

(3,200 miles) and produce a pressure and temperature profile of the atmosphere that settled the competition between the two atmospheric models.

As it turned out, neither model was entirely correct, although the 86-degree-Kelvin temperature in Danielson and Caldwell's version was only 9 degrees too low. The actual surface pressure of Titan is about 1.5 bars, or 50 percent greater than on Earth. Because Titan's gravity is lower than Earth's, however, the amount of gas present in Titan's atmosphere, per unit of surface area, is about ten times as great as in the terrestrial atmosphere. Hunten's prediction for the dominant atmospheric constituent was correct: Nitrogen makes up an estimated 82 percent to 94 percent of the atmosphere, followed by methane and perhaps argon. Complex organic compounds such as hydrogen cyanide and cyanogen are also present in abundance.

The compounds and long-chain polymers that make up Titan's obscuring photochemical haze are believed to be produced in reactions touched off by incoming solar energy. As the reactions cause the particles to grow in size, they should fall out of the clouds and pile up on the surface in what may be 109.4-yard-thick layers of what some scientists refer to as "organic goo." Laboratory simulations of the chemistry of Titan's atmosphere have produced a lengthy inventory of organic compounds of the kind thought to have been created on Earth more than four billion years ago. In the view of many researchers, Titan is a terrestrial world in a deep freeze. Only the low temperature prevents the formation of more complicated molecules and, perhaps, life itself. As Voyager scientist Tobias Owen has put it, "We are not talking about life originating on Titan, but we are talking about the first chemical steps toward life." Even in the absence of life, the Titanian surface must be a marvelous place. There may be vast lakes and seas of liquid ethane that encircle continents of ice, rock, and solid carbon dioxide (such land masses

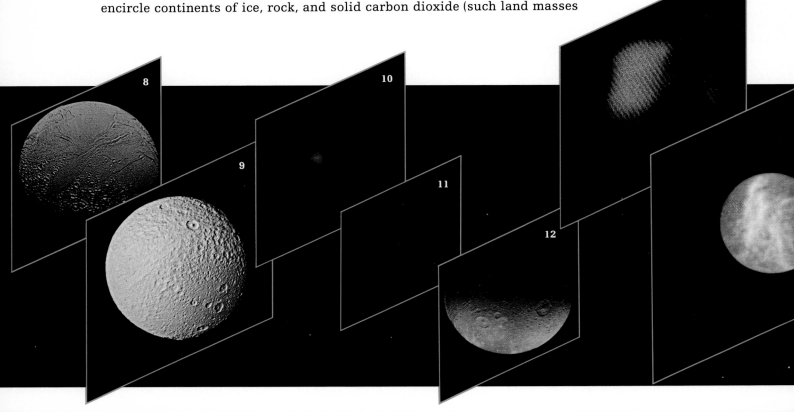

have been indicated by Earth-based radar studies). Some scientists have speculated about the possible presence of active volcanoes. The first radar images of the Titanian surface from the Cassini mission will find an eager audience awaiting them on Earth.

SURPRISES AT URANUS

Before *Voyager 2*'s Uranian encounter in January 1986, very little was known about the planet—and almost nothing about its moons. The five satellites discovered from Earth, mere pinpoints of light in the best telescopes, were thought to be in the medium-size range, comparable to Saturn's major moons. In an environment much colder than that of Jupiter or Saturn, with little energy to drive geologic processes, the Uranian moons were expected to be dead, crater-pocked ice worlds perhaps similar to Mimas. These expectations met with the same fate as presumptions earlier in the mission. During the Uranus encounter, Laurence Soderblom, deputy leader of the Voyager Imaging Team, said in wonderment, "We've had the same terrestrial, conservative attitude at Jupiter and Saturn, and we're being surprised again at Uranus. I wonder if we'll learn by the time we get to Neptune."

Among other things, Voyager found two new Uranian rings, bringing the planet's total to eleven, and ten new moons, ranging in size from fifteen miles to ninety-six miles across. Almost all of the new moons are situated between the outermost ring and the innermost of the five largest satellites. They are misshapen and dark, and spectral studies suggest they may be composed mainly of water ice. Possibly, their darkness is the result of exposure to charged particles trapped in the Uranian magnetic field, or perhaps it is due to contamination by carbonaceous material such as is present in the rings.

The five larger moons (proceeding outward from the planet: Miranda, Ariel,

Saturn's icy outer moons *(below)* display a notable geology. The smooth areas on 310-mile-wide **Enceladus** *(8)* may have been resurfaced by water volcanism. **Tethys** *(9)*, twice the size of Enceladus, is densely cratered in places but has also seen some resurfacing. Orbiting about 60 degrees ahead of and behind Tethys are its tiny coorbitals, **Telesto** *(10)* and **Calypso** *(11)*. **Dione** *(12)*, akin to Tethys in size but with no sign of geologic activity, orbits about 60 degrees behind its small partner **Helene** *(13)*. **Rhea** *(14)*, 951 miles wide, juxtaposes streaks of ice with cratered areas of great antiquity. Very little is known about the surface of 3,200-mile-wide **Titan** *(15)*, shrouded by a dense atmosphere. **Hyperion** *(16)*, about one-twentieth the size of Titan, is marked by repeated impacts. The motley look of 900-mile-wide **Iapetus** *(17)* may in part be the result of a comet that sprayed one side of the moon with its ruddy remains. Outermost **Phoebe** *(18)*, 137 miles across, is a foundling; its relative lack of surface ice and its irregular orbit suggest that it formed elsewhere in the Solar System and later fell into Saturn's gravitational embrace.

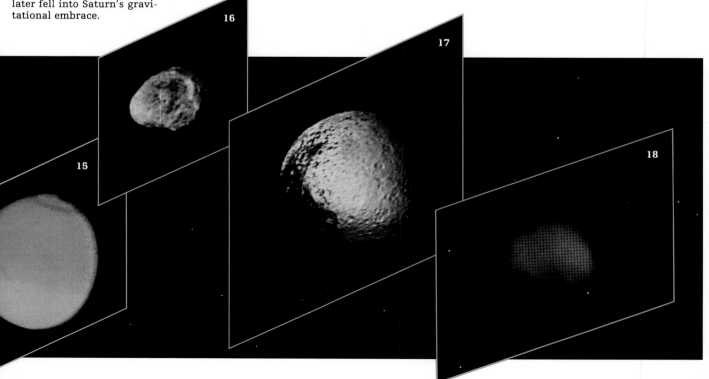

Umbriel, Titania, and Oberon) are also quite dark—more "dirty ice balls," in the parlance of planetary science. Their densities imply a composition that is about half ice and half rock. Oberon fitted the scientific preconceptions. It is a vast field of craters and little else. Titania, however, revealed rift valleys several hundred miles in length and evidence of early episodes of resurfacing by a dark material, possibly methane or water ice. Umbriel, the middle member of the five, is the darkest and deadest of the Uranian moons; except for a feature nicknamed the "fluorescent Cheerio," its surface is more uniformly dark than those of Oberon and Titania, which show numerous splashes of bright water ice thrown up from the interior by impacts.

Ariel, nearly the same size as Umbriel and next in toward the planet, presents a startling contrast to its neighbor. The brightest of the Uranian satellites, reflecting 40 percent of the light that falls on it, its surface has apparently been recoated with fresh water ice many times. In addition, Ariel displays a variety of geologic terrains, including faults, rifts, grooves, ridges, and relatively craterless plains. Scientists suspect that ice volcanism has played a major role in shaping the surface of Ariel. Erupting through vents in valley floors, mixtures of water ice, methane, and ammonia may have flowed across the surface much like terrestrial glaciers, scouring the old terrain and covering it with a fresh, smooth coating. The process seems similar to what might have occurred on Enceladus, but again, the source of the necessary energy remains uncertain. Although the moon's present orbit makes a tidal resonance with neighboring Umbriel impossible now, such regular tugging may have occurred at some time in the past.

Seen here in enhanced color, the thick haze of Titan's gaseous atmosphere completely obscures the moon's surface. Below the smog, liquid ethane—condensed from the ethane that forms through the action of sunlight on atmospheric methane—might have accumulated to form vast, icy seas.

Through the luck of the draw, *Voyager 2* made the closest approach of its twelve-year mission (22,400 miles) to tiny Miranda, just 300 miles in diameter. The choice of trajectory had nothing to do with Miranda's scientific desirability, which was thought to be minimal, but was dictated by the necessity of acquiring a gravitational assist from Uranus to get the spacecraft to Neptune; Miranda happened to be the only moon near the aiming point. The flyby was close enough to make images with a surface resolution of less than .6 mile; scientists expected to see Miranda's blandness in great detail.

Instead, the pictures transmitted back to Earth portrayed "a satellite designed by a committee," as Brad Smith described it. Prominent on the surface were three huge structures consisting of uplifted central plateaus surrounded by steep terraces. One sheer cliff was nearly nine miles high, about ten times higher than the rim of the Grand Canyon. Elsewhere, rolling cratered uplands suddenly gave way to vast systems of faults and grooves and jumbled terrain. Larry Soderblom described the little moon as a "bizarre hybrid of the valleys and layered deposits on Mars, combined with the grooved terrain on Ganymede and the compressional faults on Mercury." It was as though some mischievous god had assembled scraps of every exotic terrain elsewhere in the Solar System and deposited them all on Miranda.

Geologists gaped in disbelief, then scrambled to come up with a logical explanation for what they saw. There was general agreement that the uplifted plateaus, called coronae for their resemblance to royal crowns, must consist of ice that is less dense than the surrounding ice-rock mix. One theory proposes that Miranda is a case of arrested development. Early in its history, goes this notion, the little moon ran out of internal heat before its constituents were fully differentiated—that is, before all the heavy material could sink to the center. If the differentiation had been able to continue uninterrupted, the entire surface would be covered with the icy material that is now confined to the coronae. Another theory suggests that the process of differentiation was interrupted by something considerably more dramatic than a local energy crisis. Eugene Shoemaker of the U.S. Geological Survey, who helped reconstruct the violent history of Earth's moon, has suggested that Miranda might have been completely shattered—repeatedly—by catastrophic impacts.

An ebullient field geologist, Shoemaker has spent years smashing rocks with his hammer. In 1982, he became convinced that nature has been doing much the same thing with the moons of the outer Solar System. His idea gained attention during the Saturn encounters, when the Deathstar crater observed by *Voyager 1* on Mimas raised the possibility of world-destroying collisions. Miranda seemed to be Exhibit A. Early in its history, a partially differentiated Miranda might have been blasted apart by the impact of a large comet. But rather than flying off in all directions, the pieces of the proto-Miranda would have stayed in roughly the same orbit and, over time, gravitationally reaccreted. Each time Miranda pulled itself together, however, another comet came along and smashed it again. According to Shoemaker, Miranda could have been destroyed and reassembled five times or more, a violent process that produced the chaotic jumble of terrains observed today.

One problem with Shoemaker's thesis is that the repeated poundings should have broken Miranda into innumerable little pieces rather than the big chunks suggested by the jigsaw surface. Another problem is that the coronae appear to be relatively young; the near absence of cratering in their centers and the sharpness of their walls suggest that they are perhaps no more than a few hundred million years old. But uncertainties abound; Larry Soderblom describes the age of Miranda as "young, plus or minus old." With no future Uranus missions contemplated, geologists will have to unravel the mystery of Miranda from their archive of Voyager images. If the moon is indeed a jigsaw puzzle, scientists can only hope that none of the pieces are missing.

THE LAST PICTURE SHOW

In late August 1989—twelve years and nearly three billion miles after it was launched—*Voyager 2* approached Neptune for what was to be its last planetary encounter. Scientists and a multinational press corps gathered once again at the Jet Propulsion Laboratory to share the excitement of exploration, of seeing unknown worlds through proxy eyes. Experience had taught them to expect the unexpected, and there was already reason to think that this final flyby might turn out to be the most surprise-laden of all. Neptune did not disappoint.

The mystery of the ring arcs was at least partially solved. *Voyager 2* found four rings, all of them completely encircling the planet. But most of the material in the outermost ring was concentrated in three thirty-five-degree segments that had been observed as the arcs; the rest of the ring was simply too diffuse to be seen from Earth. Theorists were somewhat relieved to discover complete rings, but nonetheless found the ring structure, described as "strings of sausages," difficult to understand. "They are still arcs, no matter what anyone tells you," said Carolyn Porco, a planetary scientist and ring specialist at the University of Arizona.

The rings' abundance of dust is another theoretical challenge. Even if in-falling comets were a hundred times more common at Neptune than at Saturn, they still could not provide the observed amount of ring dust. The alternative is a goodly number of small moons that might replenish the supply of ring particles by breaking apart in collisions. Although no such profusion of minisatellites was detected, scientists are not ruling them out. "I think we have a lot of work to do," said Porco. "We're basically fishing around for ideas."

114

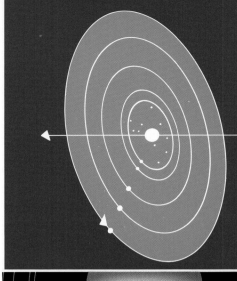

Cordelia
Ophelia
Bianca
Cressida

Desdemona

Juliet

Portia

Rosalind

Belinda

Puck

Miranda

Ariel

Umbriel

Titania

Oberon

URANUS

The Uranian satellites are the most orbitally regular in the outer Solar System. The fifteen known moons—five major ones and ten smaller ones discovered by *Voyager 2*—all lie close to the up-ended planet's equatorial plane, and their prograde paths are almost perfectly circular *(top)*. The ten small moons resemble the inner satellites of Saturn and Jupiter in that they appear to be products of collisions. But only two of the Uranian shards play a part in the planet's ring system—the shepherds Cordelia and Ophelia.

Planetary Orbit

An infrared photograph of the Uranian environment *(below)* shows the five major moons—frozen spheres coated with varying amounts of fresh ice and what is thought to be meteoritic dirt—grouped around the larger surface of the planet. Directly to the right of Uranus is outermost **Oberon,** nearly a thousand miles wide. Ranging in toward the planet are **Titania** *(bottom),* a body about the same size as Oberon; **Umbriel** *(top),* about 700 miles across; and the slightly smaller **Ariel** (just below Uranus and to the right). Directly above the planet lies its nearest major satellite, 300-mile-wide **Miranda,** whose patchwork surface *(inset, near right)* suggests that it sustained a number of shattering impacts and reaccreted haphazardly. A computer simulation *(far right)* offers a closeup view of the tortured moonscape that resulted.

Before the encounter, Neptune was known to possess only two moons, both of them odd. Tiny Nereid, just 210 miles in diameter, was discovered by Gerard Kuiper in 1949. It has the most highly eccentric orbit of any known moon, and takes 360 days to complete one trip around Neptune. Triton, slightly smaller than Earth's satellite, is the only large moon in the Solar System with a retrograde orbit around its planet. This suggested to astronomers that Triton had formed elsewhere and been captured by Neptune's gravity. Analysis of Triton's spectrum had revealed methane and nitrogen, implying that, like Titan, Triton possessed an atmosphere—although one far less dense than that of the Saturnian satellite. Preencounter speculation centered on the possibility that Triton's nitrogen might be in liquid form, and that the surface might be partially covered by a sea of liquid nitrogen.

Nereid was seen only from a distance by Voyager, and images revealed little

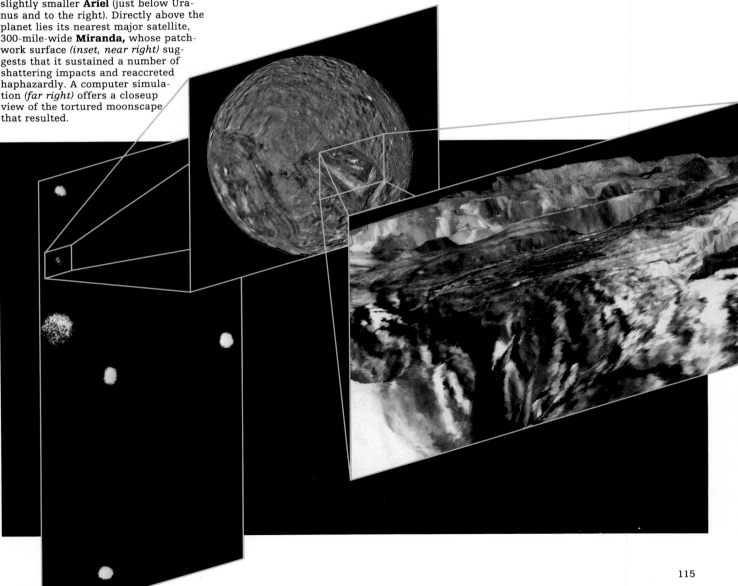

115

other than that it is very dark, with a reflectivity of just 14 percent. The spacecraft discovered six other Neptunian moons, including one, provisionally named 1989N1, that is slightly larger than Nereid. It had escaped prior detection because its orbit keeps it close to the glare of Neptune and because it is even darker than Nereid, with a reflectivity of just 6 percent. The five other new moons range from 30 to 120 miles in diameter. Although two of them orbit in the vicinity of the rings, none is properly positioned to serve as the kind of ring-shepherd theorists had expected to find.

Triton would delight them. At 3:30 a.m. on August 25, 1989, high-resolution images from the space probe appeared on the JPL television monitors and told their remarkable story. Although Triton lacked nitrogen seas, it displayed a variety of terrains, with evidence of large-scale faulting and expansive basins that seemed to be, if not liquid oceans, at least frozen lakes, probably composed of water ice. Some surface areas were very bright, covered with what seemed to be a seasonally migrating nitrogen frost. Large regions featured a curious dimpled texture that was immediately labeled the "cantaloupe terrain." The nitrogen-methane atmosphere was extremely tenuous, with a surface pressure of fourteen microbars, one seventy-thousandth that of Earth. At first, the atmosphere seemed perfectly clear, but soon scientists noted thin columns of dark material rising from the surface. A month later, these plumes, a mile to five miles high, were determined to be spewing forth from the vents of active eruptions. Like Io, Triton had been caught in the act.

Even before the plumes were spotted, it was evident that Triton's surface was one of the youngest in the Solar System. Some process was reworking this world. Larry Soderblom speculated about nitrogen volcanism at a press conference just hours after the encounter. "This is a crazy idea," he said. "It's probably wrong, but it's the best we have. We're thinking in real time." Soderblom's real-time thinking turned out to be quite good, although scientists later settled on "geysers" as a more accurate term than "volcanoes." In the absence of an internal heat source, and with no tidal heating possible as at Io, solar energy seems to be the best candidate for powering the nitrogen geysers. A subsurface layer of nitrogen gas could build up pressure under weak points on the surface—perhaps cracks in the ice produced by solar heating as the seasons slowly change on Triton. When the pressure is great enough, the gas blasts through the surface layer. "Make no mistake about it," said Brad Smith, "these events are violent."

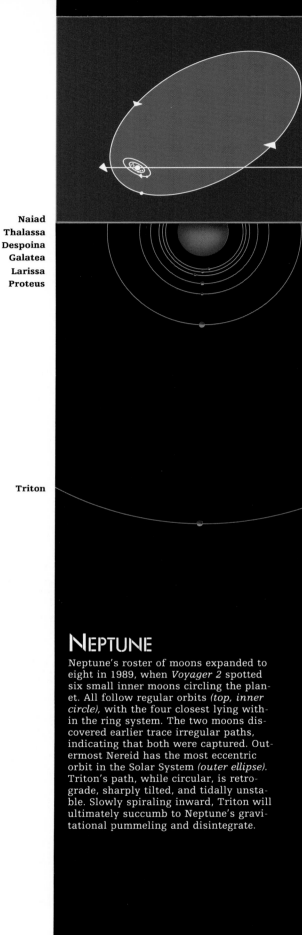

Naiad
Thalassa
Despoina
Galatea
Larissa
Proteus

Triton

Neptune

Neptune's roster of moons expanded to eight in 1989, when *Voyager 2* spotted six small inner moons circling the planet. All follow regular orbits *(top, inner circle)*, with the four closest lying within the ring system. The two moons discovered earlier trace irregular paths, indicating that both were captured. Outermost Nereid has the most eccentric orbit in the Solar System *(outer ellipse)*. Triton's path, while circular, is retrograde, sharply tilted, and tidally unstable. Slowly spiraling inward, Triton will ultimately succumb to Neptune's gravitational pummeling and disintegrate.

Nereid

A surface temperature of 38 degrees Kelvin makes Triton the coldest world ever visited. Yet it is not simply a large ice ball, like some moons of Saturn and Uranus. Triton's density is about twice that of the ice moons, implying a composition that is more rock than ice. As a captured body, it may be a sample of early conditions elsewhere in the primordial solar nebula. In fact, its temperature, density, and composition are all remarkably similar to another world: "Triton is a lot like Pluto," said Voyager project scientist Ed Stone. "And it will probably be our best look at Pluto for a long time to come."

THE UNSEEN WORLD

Ever since its discovery in 1930, Pluto has been a source of frustration for astronomers. Basic facts about its size, density, reflectivity, and constituents could not be determined with any confidence, due to Pluto's small size and great distance. But James Christy's discovery of Pluto's moon Charon in 1978 *(Chapter 1)* inaugurated a new era of research into this farthermost member of the solar family. The timing could not have been better. Like Uranus, Pluto is tipped over severely on its axis, with its north pole dipped slightly below the plane of its orbit and its equator perpendicular to it. Charon orbits above Pluto's equator. By a grand stroke of luck, during the period from 1985 to 1990, Pluto and Charon moved into an orientation that presented observers on Earth with an edge on view of Charon's orbit, an arrangement as fortuitous as the face on bull's-eye presented by Uranus. This orientation allowed ob-

Voyager 2 passed almost three million miles from **Nereid** and obtained only a fuzzy view of that dark, 210-mile-wide captive *(below)*. But the probe swept within 25,000 miles of **Triton**—a much brighter body almost eight times the diameter of Nereid—and recorded the moon in exquisite detail. In the mosaic at center, the moon's south polar cap, covered with nitrogen frost and methane ice, is to the left, and a baffling region dubbed "cantaloupe terrain" for its bumpy surface is to the right, near the equator. In a find that echoed the discovery of volcanoes on Io a decade earlier, *Voyager 2* detected telltale geyser-like plumes on Triton (black wisps in the closeup at far right), believed to be the explosive result of sunlight heating underground nitrogen deposits.

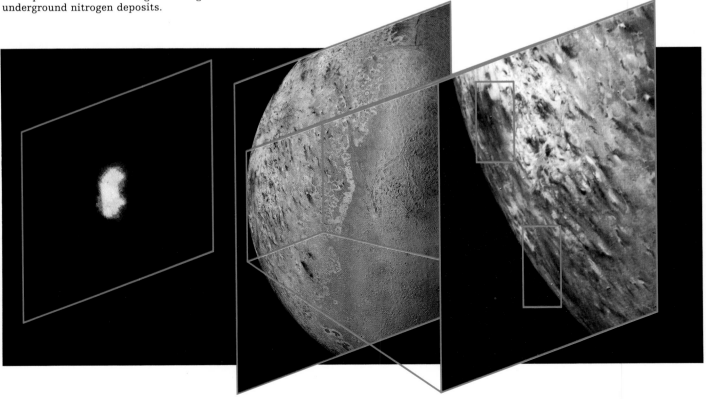

servation of the two bodies as they occulted each other *(above)*. The last time they were so disposed, General Ulysses S. Grant was laying siege to Vicksburg and Robert E. Lee was marching on Gettysburg. If Christy's discovery had come a few years later, astronomers would have missed their once-in-a-century chance to observe the revealing mutual occultations.

Pluto and Charon are more like a double planet than a planet-and-moon system. Pluto's estimated diameter is just 1,457 miles—two-thirds that of Earth's moon—while Charon's diameter is about 740 miles. For Earth, the planet-to-moon mass ratio is 81 to 1; for Pluto and Charon, it is only 10 to 1. Two consequences flow from this. For one, the mutual center of gravity about which both bodies orbit is located in space between them, rather than within the mass of the planet as is the case with all other planets and their satellites. Similarly, Pluto and Charon are in mutual tidal lock. Most large moons eternally present the same face to their planet; uniquely, Pluto also presents the same face to its moon. Moreover, Charon orbits just 12,000 miles above the surface of Pluto, making it loom large in the Plutonian sky. Earth's moon can be eclipsed by a pea held at arm's length; eclipsing Charon from Pluto would require a billiard ball.

When the planet and moon began their dance of mutual occultations, astronomers took advantage of the opportunity to detect features on Pluto.

During their 248-year trip around the Sun, Pluto and its massive satellite Charon corkscrew around a shared center of gravity in such a way that, twice in each solar orbit, they occult each other *(top)* as they are viewed from Earth. Such was the case in 1990, when the Hubble Space Telescope recorded an unprecedented double portrait *(above)* of these distant outriders.

During the five years that the occultations lasted, the gradually shifting orientations of the two bodies with respect to observers on Earth caused Charon to appear to cover up first one of Pluto's poles, then the equator, and finally the other pole. By carefully analyzing the rise and fall of the total light output of the pair, astronomers could roughly map the dark and bright areas on Pluto's surface. Its poles are bright, suggesting a fresh coating of ice, probably methane, that has frozen out of its tenuous atmosphere, a phenomenon that increases when the planet is at its farthest from the Sun. The equator is relatively dark, implying that surface ice has been darkened by radiation or that naked rock may be poking through in some regions. Overall, Pluto is a very bright body, with a reflectivity of about 47 percent, while Charon is a darker, grayer object, reflecting about 25 to 40 percent of the light reaching its surface. Charon's relative darkness may be due to its lack of an atmosphere, preventing bright frost from forming.

Because of their similarity to Triton in size, density, and general location, Pluto and Charon may have formed in the same neighborhood as the Neptunian moon; they may all have been moons of Neptune at some point, although this speculation is now deemed rather unlikely. More probable is a scenario involving the birth of all three bodies as part of a larger population of planetesimals in the outer Solar System. As the solar family evolved, most of these chunks of matter were incorporated into Neptune or Uranus (which may be tipped over as a result of being smacked by one of these moon-size bodies). Survivors might have shattered in collisions with moons or comets. Triton was finally captured by Neptune, but Pluto and Charon managed to evade the gas giants and went on to form their own lonely double-planet system. If this scenario is correct, then one or both of them may indeed closely resemble Triton. On the other hand, Triton's capture may have produced so much internal heat that it was transformed in a manner never experienced by Pluto and Charon. The only way to find out is to send a mission to them.

POST-VOYAGER MISSIONS
Because the Voyagers are headed out of the Solar System after their sixty-world reconnaissance, an entirely new mission will be needed to get to Pluto. One scheme in the planning stage at the Jet Propulsion Laboratory calls for a small, lightweight spacecraft to be launched by a Delta rocket booster in November 2001. After a gravity-assist at Jupiter in May 2006, it would conduct a flyby of Pluto and Charon in June 2015. As with most planned missions to the outer Solar System, the launch date is critical; the opportunity for a Jupiter-to-Pluto gravity-assist occurs just once every thirteen years, and if this one is missed, another Pluto flyby could not come until 2028.

Other post-Voyager missions are nearer at hand. After a six-year flight, the spacecraft for the Cassini mission to Saturn is scheduled to go into orbit around the planet in 2002 and drop a probe into the dense atmosphere of its moon Titan. Descending via a series of parachutes, the probe will return more than an hour's worth of data before a less-than-soft impact on the surface.

It is just possible that the probe will survive the landing, although no one can say whether it would hit a patch of dry ground or, emulating America's first space shots, splash down in the ethane oceans. Even as Cassini is approaching Saturn, a radar mapper aboard the spacecraft will reveal Titan's surface in much the same way that the Magellan spacecraft has begun to map the face of cloud-shrouded Venus. The orbiter will also make new and detailed images of Saturn's rings and other moons.

One probe of the outer Solar System—the Galileo mission—is already on the way toward Jupiter. It will arrive in December 1995. After dropping a package of instruments into the seething atmosphere of the gas giant, Galileo will make multiple looping orbits of the planet, passing close to each of the Galilean satellites. The spuming volcanoes of Io will be imaged at a resolution of just twenty-two yards. Many of the lingering mysteries of Europa, Ganymede, and Callisto may be cleared up.

No matter how successful Galileo and Cassini are, they will be hard-pressed to match *Voyagers 1* and *2.* The two new missions will be the Balboas and Hudsons of space exploration; but the two Voyagers were the robotic equals of Columbus and Magellan. "It will never happen again like it's happened this time," said Voyager scientist Carolyn Porco as the Neptune encounter drew to a close. "We will never approach a planet in the same state of ignorance and innocence."

CIRCLES OF UNREST

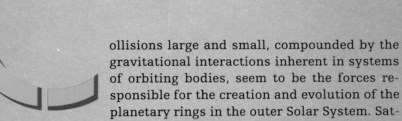

ollisions large and small, compounded by the gravitational interactions inherent in systems of orbiting bodies, seem to be the forces responsible for the creation and evolution of the planetary rings in the outer Solar System. Saturn's spectacular collection by far outshines the assorted rings of its siblings, but, as a decade's worth of Voyager missions has shown, even the modest bands of particles around Jupiter, Uranus, and Neptune yield insights and clues that help scientists understand the rings' origins and behavior.

The first lesson taught by the Voyager probes was that rings are not all alike. The Uranian bands are so narrow and dark that only twentieth-century spectroscopic techniques could discern them from Earth; the broad expanse of Saturn's rings, of course, has fascinated observers since Galileo first spied them through his telescope in 1610. Some rings contain boulders that are as large as houses; others are mere smokelike trails made up primarily of microscopic dust. One ring that is wide and diffuse may lie just inside another that is narrow and crisply defined.

Whatever their dimensions and composition, planetary rings are evidence of ongoing evolution in the Solar System's outer reaches. Ring particles collide with one another almost continuously, and photons of light knock particles out of their orbits, causing them to plummet to a fiery end in the planetary atmosphere below. But even as the rings are depleted of the bits that compose them, constant micrometeoroid bombardment blows a fresh supply off the surfaces of small moons, to enter the gravitational dance around the large planets that lie beyond Mars.

PARTICLES OF DUST IN MOTION

Sometimes the most striking effects in planetary rings are produced by the smallest players. Because the dust motes exploding off the surfaces of ring particles struck by micrometeoroids are more sensitive than their larger counterparts to relatively weak forces, their behavior often has dramatic repercussions. Saturn's broad B ring, for example, exhibits radial markings, known as spokes, that appear in only minutes and then disappear in a few tens of hours. Scientists theorize that electrically charged dust particles are influenced by the planet's magnetic field lines *(box, far right)*, and that because fine particles scatter light differently from larger particles, they show up as dark or light streaks in the rings, depending on the direction of view.

In other parts of the ring system, various kinds of drag cause small ring particles to drift down toward their planet. For instance, dust motes colliding with charged atoms in the magnetosphere lose energy, and they drop steadily to lower and lower orbits. In addition to being subject to this so-called plasma drag, small ring particles are also at the mercy of head-on collisions with photons of sunlight, a phenomenon known as Poynting-Robertson drag *(above, near right)*. Both forms of drag can result in the particles' ultimately being pulled down into the atmosphere and vaporized.

Poynting-Robertson drag. As microscopic dust particles orbit in a planetary ring, photons of sunlight tend to strike their leading edges. The resulting loss of energy causes the particles to drop to lower orbits until they are burned up in the planet's atmosphere.

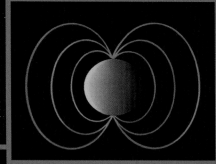

Spokes. One theory to explain the creation of spokelike markings in Saturn's B ring, illustrated at left, involves magnetic field lines in the planet's magnetosphere *(above)*. According to this hypothesis, a meteor collision with a large ring particle produces a cloud of plasma, or ionized atoms, which shoots out perpendicular to the field lines—that is, radially across the ring. The plasma cloud charges microscopic dust particles on the surfaces of larger particles lying in its path. The dust, now influenced by electromagnetic forces, rises off the surfaces of the larger particles in an effect similar to that of hair-raising static electricity on Earth. The light-scattering effect of these narrow veils of dust creates distinctive markings when viewed from a distance. Briefly held aloft in the orbital dance, the spokes start out as straight lines but gradually diffuse as particles at the leading and trailing edges orbit at different rates. Eventually, particle collisions cause the motes to drop back onto the surfaces of their parent particles.

As shown above, a ring particle *(white)* in a 2:1 resonance with a moon *(gray)* orbits once *(red line)* while the moon travels through half an orbit. On every other orbit, the particle is tugged by the moon until it shifts out of its old path *(dashed line)* into a new elliptical one *(blue line)*. Because nearby particles are affected at slightly different times, the various orbits rotate relative to each other *(below)*, increasing the particle density and producing a spiral pattern *(right)*.

Generating Spiral Density Waves

Among the more complex features of planetary rings are two phenomena known as spiral density waves *(left)* and their three-dimensional counterparts, spiral bending waves *(below)*. Both patterns are produced by repeated gravitational interactions between ring particles and nearby moons. Like all orbiting bodies, particles and moons travel at specific velocities that depend on their distance from their planet. When the orbital periods of two bodies happen to be simple ratios of each other, the bodies are said to be in resonance. For example, a ring particle that completes two orbits of its planet in exactly the time it takes a moon to orbit once is in a 2:1 resonant orbit with the moon *(diagram, far left)*.

Whenever the particle and moon are aligned—once every other orbit for a particle in 2:1 resonance—the moon exerts an extrastrong gravitational pull on the particle. Over time, the repeated tug at the same point in the particle's orbit will cause its path to become more elliptical. Particles near resonance points will also be pulled into somewhat elliptical tracks, and where these new orbits converge, gravitational attraction between particles will cause them to bunch up. The combined gravity of the accumulation of particles will in turn tug on the rest of the ring, causing particles elsewhere to clump up in some places and to thin out in others. These regions of condensation and rarefaction produce a wave perturbation that moves through the ring, creating a spiral pattern especially visible in Saturn's outer A ring.

The two-armed spiral density wave at left is the product of a 2:1 resonance between a cluster of ring particles and a moon orbiting in the same plane. But when a moon happens to follow an orbital path inclined to the plane of the ring *(blue line, below)*, its repeated tugs elevate the resonant particles into an orbit *(red line)* inclined to the ring plane, generating a phenomenon known as a spiral bending wave. A handful of bending waves have been discovered in Saturn's ring system.

A pair of shepherding moons that have largely cleared their paths of ring particles force the particles between them into a narrow, sharp-edged band. As the inner moon, obeying orbital mechanics, catches up with and passes slower-moving particles farther out, its gravitational attraction pulls the particles in their direction of motion, transferring angular momentum that boosts them to an even higher orbit. Conversely, the outer moon, traveling more slowly than the ring particles, exerts gravitational drag that robs passing particles of energy so that they drop to lower orbits.

Scientists witnessing the medley of ring photographs beamed back by the Voyager probes have often been hard-put to explain some of the features they saw. The narrow rings of Uranus, for example, most of them less than eight miles across, seemed to defy the laws of physics that require colliding particles to disperse *(pages 104-105)*. After considering and then discarding an elaborate resonance scheme involving five visible moons to confine the ring particles to their narrow paths, astronomers turned to the idea that unseen moons might be acting as gravitational "shepherds," with ring particles as the sheep.

The notion may explain not only the narrow rings at Uranus but also some of the filaments mixed in with Sat-urn's broader bands. Shepherding moons work in pairs, with one inside the ring orbit, one outside. The inner satellite, through a transfer of angular momentum, pushes the ring particles outward while the outer satellite pushes them inward *(far left)*. The idea was confirmed when *Voyager 1* discovered two small moons shepherding Saturn's F ring.

Small moons are also suspected of creating gravitational disturbances that create odd wiggles and wavy edges, visible only in images made at extremely close range. These scalloped features are actually the initial phase of the shepherding process, and appear only in short portions of rings and only when a moon is passing by. As the moon moves on, and particle collisions dampen the wave pattern, the disturbed ring particles settle back into their normal orbital paths until the moon's next passage.

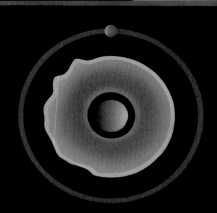

Among the products of the gravitational interaction of moons and ring particles are rings with scalloped edges. A moon orbiting inside a ring *(right)* tugs slower-moving ring particles toward it and moves on, leaving a train of bumps in its wake that eventually damp down. A similar effect occurs when a moon orbits just outside of a ring *(left)*. Because the moon is traveling more slowly than the ring particles, the ripple effect travels through the ring ahead of the moon.

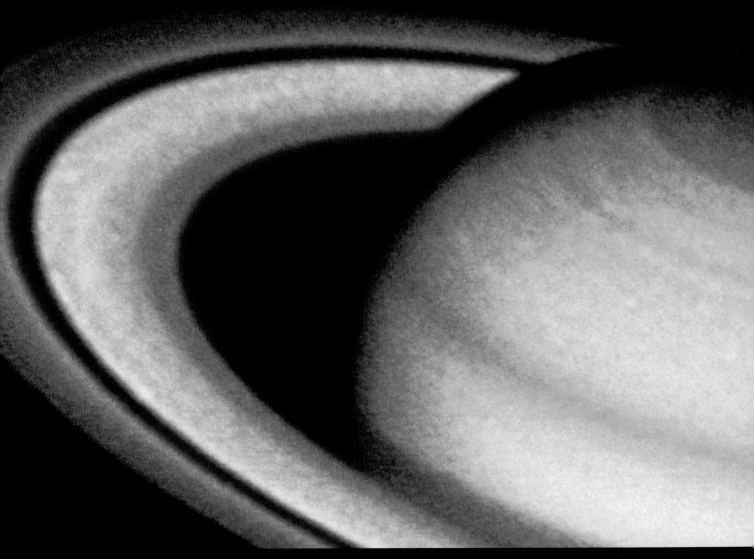

THE SPECTACLE AT SATURN

Saturn's vast system of rings is by far the most complex of any in the Solar System and serves as a grand laboratory for ring research. The photograph shown here, taken by the Hubble Space Telescope, is one of the most detailed images ever made from the vicinity of Earth, clearly delineating the Encke Division that segments the planet's A ring. But it took closeup examination by the Voyager probes to reveal the truly ornate nature of Saturn's rings. The probes showed that the original trio of broad bands discerned from Earth actually consists of innumerable narrow features and wispy ringlets in regions that were thought to be empty. Despite the unprecedented views supplied by Voyager, however, the Saturnian system still guards many of its secrets. For example, scientists do not yet know what, exactly, the ring particles are made of. Although the orbiting boulders and bits are primarily ice, various other materials must lie within or on the surface of the particles to give the rings their reddish brown tinge in visible light.

One of the more baffling of Saturn's ring features is the braided segment of the F ring, a narrow ribbon that lies 2,175 miles beyond the A ring. The eccentric orbits of two small shepherd moons—Pandora and Prometheus—are believed to cause the kinks, which orbit the planet as a fixed feature of the F ring. The changing gravitational interaction between the faster-moving inner moon and its slower partner may cause the ring particles to follow paths that twist in different ways.

In the Earth-based infrared image of Uranus shown here, blue indicates the greater reflectivity of the planet compared with the nearly invisible rings *(red)*. The fuzzy red circle represents blurred infrared light reflected from the Epsilon ring, the most substantial Uranian ring and the one that lies farthest from the planet. As shown in the illustration below, however, Uranus possesses at least ten other rings, many of them just over a mile in width.

1986U2R

6

5

4

Alpha

Beta

Eta

Gamma

Delta

1986U1R

Epsilon

THE DARK, NARROW BANDS OF URANUS

Unlike Saturn's broad bands, long a favorite subject of amateur astronomers, Uranus's rings were unknown before 1977, when several groups of observers caught an unexpected series of discrete dips in light level caused by the rings as the planet occulted a distant star. Not only are the Uranian rings all exceedingly narrow—the broadest, Epsilon, is only some sixty miles across at its widest—but they are also charcoal-black, reflecting scarcely one percent of the sunlight that reaches them. The low reflectivity of the ring particles suggests a high carbon content similar to that of many asteroids, especially those in the outer part of the main asteroid belt.

Before *Voyager 2* reached Uranus in 1986, the planet was thought to have nine rings. The spacecraft discovered at least two additional rings, sandwiched between larger ones on the outer edge of the system. Voyager also revealed the existence of a number of moonlets, most of which lie beyond the ring system, as well as more than a hundred diffuse, dusty bands.

The origin of the rings remains uncertain, although one theory suggests that the particles are the result of the destruction of small moons by cometary meteoroids. Whatever their genesis, the dusty rings are believed to have formed no more than 10 million to 100 million years ago; were they much older, the ring particles would have been dragged to fiery oblivion in the planet's atmosphere.

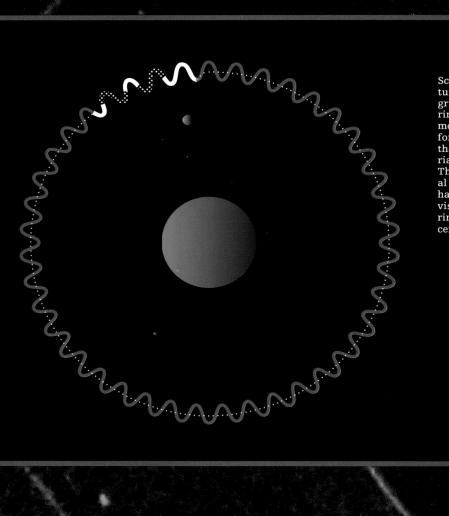

Scientists seeking to explain Neptune's arcs theorize that a complex gravitational interaction between ring particles and a Neptunian moonlet generates a complicated, forty-two-lobed resonance pattern that moves through the ring material at the same rate as the moon. The arcs are regions of gravitational equilibrium where ring particles have bunched up enough to become visible; particles in the rest of the ring remain too diffuse to be discerned from Earth.

seemed to indicate that Neptune was encircled by three arcs that appeared to be incomplete rings, some scientists hypothesized that the arcs were simply the densest—and thus most visible—parts of what were in fact normal rings.

In its long-awaited report, *Voyager 2* showed that the arcs were segments of not three rings but one. Moreover, the probe also revealed that inside the orbit of this narrow outermost ring were another thin strand and two diffuse inner bands—a sheet of smoke-like material known as the Plateau, and a band made of fine dust that may extend all the way down to Neptune's cloud tops.

Neptune's two narrow rings, orbiting at distances of 32,934 miles and 39,148 miles from the planet's center, are shown here in a composite of two *Voyager 2* photographs. (Because the spacecraft's camera had to remain open for nearly ten minutes in order to capture enough light to show the faint rings, Neptune itself was severely overexposed; another image of the planet is superimposed in the composite.) The ring arcs appear as slightly brighter regions on the outermost ring.

GLOSSARY

Accretion theory: a theory postulating that the regular satellites and the rings of the large gas planets in the Solar System formed along with the planets by coalescing out of the surrounding nebulae.

Angular momentum: a measure of an object's inertia, or state of motion, about an axis of rotation.

Anorthosite: a light-colored, crystalline igneous rock found in the lunar highlands.

Aphelion: the point in the orbit of a planet or comet when it is farthest from the Sun. *See* Perihelion.

Apocenter: the point in an orbit farthest from the gravitational center of that orbit. *See* Pericenter.

Asteroid: any small, rocky, airless body that orbits a star. Three main types have been identified in the Solar System: carbonaceous, or C-type; silicaceous, or S-type; and metallic, or M-type.

Atmosphere: a gaseous shell surrounding a planet or other body.

Bar: a unit of air pressure equal to about 14.5 pounds per square inch. The millibar, one one-thousandth of a bar, is a more commonly used measure.

Basalt: a dark, close-grained igneous rock, formed by the hardening of lava. Most volcanic rocks are basalts.

Basin: a large, low-lying area on the surface of a planet or moon, often a crater.

Bending wave: in a planetary ring, a corrugated wave pattern caused by the gravitational pull of a moon alternately orbiting above and below the ring plane.

Capture hypothesis: a theory that explains the compositional and structural differences between a planet and its moon by suggesting that the moon formed in another part of the Solar System and was later drawn in, or captured, by the planet's gravitation.

Centrifugal force: the apparent outward force felt by a body rotating about an axis.

Charged particle: an elementary particle having an electric or magnetic field that influences its interaction with other particles. For example, electrons carry a negative charge, protons a positive charge.

Coaccretion theory: a theory suggesting that the Moon formed along with the proto-Earth from a primordial nebula of gas and dust.

Comet: an asteroid-size body of dusty ice that travels in an elongated orbit around the Sun.

Coronae: the enormous raised plateau features found on Uranus's moon Miranda, so named because their roughly circular shape suggests a crown, or corona.

C-type asteroid: a class of asteroids characterized by surfaces of carbon-rich clay. C-type bodies make up 75 percent of the objects found in the main asteroid belt.

Density: a relation of the mass of a body to its volume; in a given volume, the larger the mass, the greater the density.

Density wave: a moving pattern of compression and rarefaction. Density waves are thought to produce the grooved structure of Saturn's rings.

Doppler shift: a change in the wavelength and frequency of sound or electromagnetic radiation, caused by the motion of the emitter, the observer, or both.

Eclipse: the obscuration of light from a celestial body as it passes through the shadow of another body.

Ecliptic: the plane described by Earth's orbit around the Sun; the apparent annual path of the Sun in the sky.

Ejecta: material that is displaced by the violent impact of an object or by volcanic forces.

Electromagnetism: the phenomenon associated with electrically charged particles in motion, including electric and magnetic fields and the forces they exert. Electromagnetism does not affect neutral particles.

Ethane: a gaseous hydrocarbon having two carbon and six hydrogen atoms per molecule thought to be present in liquid form on Saturn's moon Titan.

Fault: a region of weakness in the crust of a planet or moon. Differential motion on either side of faults builds up enormous tension, eventually causing slippage or earthquakes.

Fission hypothesis: a theory proposing that the Moon formed from matter torn from the rapidly spinning proto-Earth by the Sun's gravity.

Galilean satellites: the four satellites of Jupiter discovered by Galileo: Io, Europa, Ganymede, and Callisto.

Giant impact theory: a theory proposing that the Moon was formed when a large planetesimal struck the young Earth, ejecting material that coalesced into an orbiting body.

Gravity: the force responsible for the mutual attraction of separate masses.

Hydrocarbon: one of the large number of organic compounds made up exclusively of hydrogen and carbon atoms.

Hydrogen: the lightest and most common element in the universe.

Ice volcano: a volcano that ejects ice, water, and ammonia instead of magma. Such volcanoes have been detected on Saturn's moon Enceladus and Uranus's moon Ariel.

Infrared: a band of electromagnetic radiation with a lower frequency and a longer wavelength than red light.

Irregular satellite: a satellite having one or more of the following characteristics: highly elliptical orbit, significantly inclined orbit, or retrograde orbit, according to a classification system devised by Gerard Kuiper. *See* Regular satellite.

Kelvin: the name given to the temperature scale in which zero denotes absolute zero (-273 degrees Celsius) and a unit of temperature, called a Kelvin, equals one Celsius degree.

Limb: the apparent edge of a celestial body as it is seen in the sky.

Magma: molten rock formed beneath the surface of a planet or a moon.

Mare, *pl.* **maria:** an expanse of dark, level material on the surface of a planet or satellite.

Meteorite: the recovered fragment of a rocky or metallic body that has survived its transit through a planet's atmosphere. The weight of a meteorite may range from just a few ounces to nearly a hundred tons. A micrometeorite is a particle smaller than .1 millimeter in diameter with a mass of less than 10^{-6} gram.

Methane: a compound of carbon and hydrogen having one carbon and four hydrogen atoms per molecule, found in the atmosphere of Saturn's moon Titan and thought to be present as methane ice on other moon surfaces.

Millibar: *see* Bar.

Moon: one of a planet's natural satellites, generally no smaller than ten miles in diameter.

Nebula: a diffuse cloud of gas and dust. For example, the solar nebula is the diffuse primordial matter believed to have surrounded the young Sun and to have formed, by accretion, the nine planets as well as all asteroids, meteoroids, and comets.

Nebular hypothesis: the theory describing the process by which the bodies of the Solar System formed from a primordial nebula.

Occultation: an event in which one celestial body passes in front of another, partially or totally obscuring it.

Opposition: the alignment of two celestial bodies on opposite sides of the sky as seen from Earth. A perfect opposition occurs when a planet is also at its closest approach to Earth, at its best point for observation.

Orbit: the path of an object revolving around another object.

Orbital period: the length of time it takes for a satellite to complete one orbit around its primary.

Particle: the smallest component of any class of matter; for example, the elementary particles within an atom (such as electrons, protons, and neutrons); or the smallest forms of solid matter in space (interplanetary and interstellar dust particles).

Pericenter: the point in an orbit closest to the gravitational center of that orbit. *See* Apocenter.

Perihelion: the point in the orbit of a planet or a comet when it is closest to the Sun. *See* Aphelion.

Period: *see* Orbital period.

Phase: one of the recurring appearances of a celestial body as viewed from Earth. As an object such as the Moon or a planet moves along in its orbit, the amount of its surface visible from Earth increases (waxes) and decreases (wanes) at a regular, periodic rate.

Planet: a large, nonstellar body that orbits a star and shines only with reflected light.

Planetesimal: in astronomical theory, a small, primitive body that orbited the Sun in the solar nebula, gaining mass through random collisions with other orbiting bodies until it eventually became a full-scale planet. The term derives from "infinitesimal planet."

Plasma: a gas of ionized particles, in contrast to ordinary gases, which are electrically neutral. Plasmas are sensitive to electrical and magnetic fields and are considered to be a fourth state of matter, along with ordinary gases, liquids, and solids.

Polymer: a heavy chemical compound characterized by the presence of repeating structural units of one or more types.

Primary: the object in a system of orbiting bodies closest to the center of mass of that system; the most massive object of the system. For example, Mars is the primary of Phobos and Deimos.

Probe: an automated, crewless spacecraft used to gather information or perform experiments in space or on extraterrestrial surfaces and transmit its findings back to Earth.

Prograde motion: orbital movement by a moon or planet in the same direction as that of the primary's rotation.

Radioactive decay: the spontaneous breakdown of the unstable nucleus of certain elements by the release of subatomic particles and heat.

Regolith: a layer of fragmented rocky debris that forms the surface of some planets and many moons. It is produced by collisions among orbiting bodies.

Regular satellite: a moon having a circular, equatorial, prograde orbit, according to a classification system devised by Gerard Kuiper. *See* Irregular satellite.

Retrograde motion: orbital movement by a moon or planet in a direction opposite that of the primary's rotation.

Ring: a belt of diffuse matter encircling a planet. The bodies making up the ring range in size from dust particles to boulders.

Ring arc: a segment of a Neptunian ring, made more noticeable than other sections by an aggregation of ring material along the arc.

Roche limit: the boundary zone within which a planet's tidal forces both prevent the formation of moons and may destroy orbiting moons that stray inside the limit; named after its discoverer, French astronomer Édouard Roche.

Rotation: the turning of a celestial body about its axis.

Rotation period: the time it takes a celestial body to complete one rotation about its axis.

Satellite: any body, natural or artificial, in orbit around a planet.

Schröter's Rule: a relationship stating that the volume of material mounded in a crater's outer walls is equal to the volume of the depression the walls encircle.

Secular acceleration: an increase in the orbital speed of a satellite as it draws closer to its primary.

Selenography: the study of the Moon's physical features.

Shard: a fragment of a larger celestial body shattered by collision. Shards range from boulder-size objects to small moons.

Shepherd moons: small moons, sometimes paired, that gravitationally influence the orbits of particles in some planetary rings.

Silicate: any of the largest and most common class of minerals, based on the nonmetallic element silicon. Nearly all rock-forming minerals are silicates, consisting of metal combined with silicon and oxygen.

Spectral line: a bright or dark band in an astronomical spectrum that is produced by atoms as they absorb or emit light.

Spectrometer: a spectroscope that has been fitted with scales to measure the position of various spectral lines.

Spectroscope: any of the various instruments used for the direct observation of a spectrum.

Subatomic particle: *see* Particle.

Synchronous orbit distance: the distance from a planet at which a moon will complete one orbit in the time it takes the planet to rotate on its axis. The extent of the limit is linked to the rotation rate and changes as the rotation rate changes.

Tidal bulge: a bulging of the surface of a celestial body caused by the gravitational pull of another nearby body.

Tidal flexing: a pattern of recurrent stresses caused by the gravitational pull of one or more celestial bodies on another.

Tidal force: the internal force generated in a body by gravitational interactions with another body.

Ultraviolet: a band of electromagnetic radiation that has a higher frequency and shorter wavelength than blue light.

BIBLIOGRAPHY

Books

Abell, George O., David Morrison, and Sidney C. Wolff. *Exploration of the Universe* (5th ed.). Philadelphia: Saunders College, 1987.

Baldwin, Ralph B. *The Face of the Moon.* Chicago: University of Chicago Press, 1949.

Baugher, Joseph F. *The Space-Age Solar System.* New York: John Wiley & Sons, 1988.

Beatty, J. Kelly, and Andrew Chaikin (eds.). *The New Solar System* (3d ed.). Cambridge, Mass.: Sky Publishing, 1990.

Black, David C., and Mildred Shapley Matthews (eds.). *Protostars & Planets II.* Tucson: University of Arizona Press, 1985.

Both, Ernst E. *A History of Lunar Studies.* Buffalo, N.Y.: Buffalo Museum of Science, 1962.

Burgess, Eric. *Uranus and Neptune: The Distant Giants.* New York: Columbia University Press, 1988.

Burke, John G. *Cosmic Debris: Meteorites in History.* Berkeley: University of California Press, 1986.

Burns, Joseph A., and Mildred Shapley Matthews (eds.). *Satellites.* Tucson: University of Arizona Press, 1986.

Cadogan, Peter H. *The Moon—Our Sister Planet.* Cambridge, England: Cambridge University Press, 1981.

Chapman, Clark R. *Planets of Rock and Ice: From Mercury to the Moons of Saturn.* New York: Charles Scribner's Sons, 1982.

Considine, Douglas M., and Glenn D. Considine (eds.). *Van Nostrand's Scientific Encyclopedia* (7th ed.). New York: Van Nostrand Reinhold, 1989.

Couper, Heather, and Nigel Henbest. *New Worlds: In Search of the Planets.* Reading, Mass.: Addison-Wesley, 1986.

Elliot, James, and Richard Kerr. *Rings: Discoveries from Galileo to Voyager.* Cambridge: Massachusetts Institute of Technology Press, 1984.

The Far Planets (Voyage Through the Universe series). Alexandria, Va.: Time-Life Books, 1990.

French, Bevan M. *The Moon Book.* New York: Penguin Books, 1977.

Gehrels, Tom, and Mildred Shapley Matthews (eds.). *Saturn.* Tucson: University of Arizona Press, 1984.

Gillispie, Charles Coulston (ed.). *Dictionary of Scientific Biography* (Vol. 11). New York: Charles Scribner's Sons, 1980.

Goldberg, Leo, David Layzer, and John G. Phillips (eds.). *Annual Review of Astronomy and Astrophysics.* Palo Alto, Calif.: Annual Reviews, 1968.

Green, Jack, and Nicholas M. Short (eds.). *Volcanic Landforms and Surface Features: A Photographic Atlas and Glossary.* New York: Springer-Verlag, 1971.

Greenberg, Richard, and Andre Brahic (eds.). *Planetary Rings.* Tucson: University of Arizona Press, 1984.

Hall, Angelo. *An Astronomer's Wife: The Biography of Angeline Hall.* Baltimore, Md.: Nunn & Company, 1908.

Hallion, Richard P., and Tom D. Crouch (eds.). *Apollo: Ten Years since Tranquillity Base.* Washington, D.C.: National Air and Space Museum, 1979.

Hanle, Paul A., and Von Del Chamberlain (eds.). *Space Science Comes of Age: Perspectives in the History of the Space Sciences.* Washington, D.C.: National Air and Space Museum, 1981.

Hartmann, William K. *Moons and Planets* (2d ed.). Belmont, Calif.: Wadsworth, 1983.

Hartmann, William K., and Odell Raper. *The New Mars: The Discoveries of Mariner 9.* Washington, D.C.: National Aeronautics and Space Administration, 1974.

Hartmann, W. K., R. J. Phillips, and G. J. Taylor (eds.). *Origin of the Moon.* Houston, Tex.: Lunar & Planetary Institute, 1986.

Herrmann, Dieter B. *The History of Astronomy from Herschel to Hertzsprung.* Translated and revised by Kevin Krisciunas. Cambridge, England: Cambridge University Press, 1984.

Hunt, Garry, and Patrick Moore. *Jupiter.* New York: Rand McNally, 1981.

Jones, Barrie William. *The Solar System.* Oxford, England: Pergamon Press, 1984.

Kaufmann, William J., III. *Universe.* New York: W. H. Freeman, 1985.

Kopal, Zdeněk, and Zdenka Kadla Mikhailov (eds.). *The Moon.* London: Academic Press, 1962.

Littman, Mark. *Planets Beyond: Discovering the Outer Solar System* (rev. ed.). New York: John Wiley & Sons, 1990.

Mars: As Viewed by Mariner 9. Washington, D.C.: National Aeronautics and Space Administration, 1974.

Marsden, B. G., and A. G. W. Cameron (eds.). *The Earth-Moon System.* New York: Plenum Press, 1966.

Masursky, Harold, G. W. Colton, and Farouk El-Baz (eds.). *Apollo over the Moon: A View from Orbit.* Washington, D.C.: National Aeronautics and Space Administration, 1978.

Melosh, H. J. *Impact Cratering: A Geologic Process.* New York: Oxford University Press, 1989.

Moore, Patrick:
The Moon. New York: Rand McNally, 1981.
New Guide to the Moon. New York: W. W. Norton, 1976.

Moore, Patrick (ed.). *The International Encyclopedia of Astronomy.* New York: Orion Books, 1987.

Morrison, David, and Tobias Owen. *The Planetary System.* Reading, Mass.: Addison-Wesley, 1987.

Morrison, David, and Jane Samz. *Voyage to Jupiter.* Washington, D.C.: National Aeronautics and Space Administration, 1980.

Motz, Lloyd, and Anneta Duveen. *Essentials of Astronomy.* New York: Columbia University Press, 1977.

The Near Planets (Voyage Through the Universe series). Alexandria, Va.: Time-Life Books, 1989.

Page, Thornton, and Lou Williams Page. *Wanderers in the Sky: The Motions of Planets and Space Probes.* New York: Macmillan, 1965.

Sagan, Carl. *Cosmos.* New York: Random House, 1980.

Schultz, Peter H. *Moon Morphology: Interpretations Based on Lunar Orbiter Photography.* Austin: University of Texas Press, 1976.

Shklovsky, I. S., and Carl Sagan. *Intelligent Life in the Universe.* Translated by Paula Fern. San Francisco: Holden-Day, 1966.

Shu, Frank H. *The Physical Universe: An Introduction to Astronomy.* Mill Valley, Calif.: University Science Books, 1982.

Snow, Theodore P. *Essentials of the Dynamic Universe: An Introduction to Astronomy* (2d ed.). St. Paul, Minn.: West, 1987.

Spitzer, Cary R. (ed.). *Viking Orbiter Views of Mars: By the Viking Orbiter Imaging Team.* Washington, D.C.: National Aeronautics and Space Administration, 1980.

Spudis, Paul D. "The Moon." In *The New Solar System* (3d ed.), edited by J. K. Beatty and Andrew Chaikin. Cambridge, Mass.: Sky Publishing, 1989.

Taylor, G. Jeffrey. *A Close Look at the Moon.* New York: Dodd, Mead, 1980.

Taylor, Stuart Ross. *Lunar Science: A Post-Apollo View: Scientific Results and Insights from the Lunar Samples.* New York: Pergamon Press, 1975.

Tombaugh, Clyde W., and Patrick Moore. *Out of the Darkness: The Planet Pluto.* Harrisburg, Pa.: Stackpole Books, 1980.

The Visible Universe (Voyage Through the Universe series). Alexandria, Va.: Time-Life Books, 1990.

Washburn, Mark:
Distant Encounters: The Exploration of Jupiter and Saturn. San Diego: Harcourt Brace Jovanovich, 1983.
Mars at Last! New York: G. P. Putnam's Sons, 1977.

Wood, John A. *The Solar System.* Englewood Cliffs, N.J.: Prentice-Hall, 1979.

Zeilik, Michael, and Elske v. P. Smith. *Introductory Astronomy and Astrophysics* (2d ed.). Philadelphia, Pa.: Saunders College, 1987.

Periodicals

Baldwin, Ralph B. "The Craters of the Moon." *Scientific American,* July 1949.

Baum, Richard, and Robert W. Smith. "Neptune's Forgotten Ring." *Sky & Telescope,* June 1989.

Beatty, J. Kelly. "Getting to Know Neptune." *Sky & Telescope,* February 1990.

Belcher, John W. "The Jupiter-Io Connection: An Alfvén Engine in Space." *Science,* October 9, 1987.

Berry, Richard. "Voyager: Science at Saturn." *Astronomy,* February 1981.

Binzel, Richard P. "Pluto." *Scientific American,* June 1990.

Boss, Alan P. "The Origin of the Moon." *Science,* January 24, 1986.

Boxer, Sarah (ed.). "A Shepherd Keeps Order on Neptune's Arc." *Discover,* March 1986.

Brown, Robert Hamilton, et al. "Energy Sources for Triton's Geyser-like Plumes." *Science,* October 19, 1990.

Brown, Robert Hamilton, and Dale P. Cruikshank. "The Moons of Uranus, Neptune and Pluto." *Scientific American,* July 1985.

Brownlee, Shannon. "The Moon's Birth." *Discover,* March 1985.

Brush, Stephen G.
"Nickel for Your Thoughts: Urey and the Origin of the Moon." *Science,* September 3, 1982.
"Theories of the Origin of the Solar System 1956-1985." *Reviews of Modern Physics,* January 1990.

Cassen, Patrick, Stanton J. Peale, and Ray T. Reynolds. "On the Comparative Evolution of Ganymede and Callisto." *Icarus,* 1980, Vol. 41, pp. 232-239.

Chaikin, Andrew. "Voyager among the Ice Worlds." *Sky & Telescope,* April 1986.

Chapman, Clark R. "More Surprises in Neptune Data." *Nature,* January 18, 1990.

Cooper, Henry S. F., Jr. "Annals of Space (The Planetary Community—Part I)." *The New Yorker,* June 11, 1990.

Cuzzi, Jeffrey N.:
"Ringed Planets: Still Mysterious—I." *Sky & Telescope,* December 1984.
"Ringed Planets: Still Mysterious—II." *Sky & Telescope,* January 1985.

Cuzzi, Jeffrey N., and Larry W. Esposito. "The Rings of Uranus." *Scientific American,* July 1987.

Dick, Steven J. "Discovering the Moons of Mars." *Sky & Telescope,* September 1988.

Dowling, Timothy. "Big, Blue: The Twin Worlds of Uranus and Neptune." *Astronomy,* October 1990.

Eberhart, J. "Five-Year Hunt Locates Saturn's 18th Moon." *Science News,* August 4, 1990.

El-Baz, Farouk. "The Moon after Apollo." *Icarus,* 1975, Vol. 25, pp. 495-537.

Elliot, James L., Edward Dunham, and Robert L. Millis. "Discovering the Rings of Uranus." *Sky & Telescope,* June 1977.

Esposito, Larry W. "The Changing Shape of Planetary Rings." *Astronomy,* September 1987.

Fisher, Arthur. "Birth of the Moon." *Popular Science,* January 1987.

French, Bevan M. "What's New on the Moon?—I." *Sky & Telescope,* March 1977.

Garwin, Laura. "Tales of a Lost Magma Ocean." *Nature,* March 2, 1989.

Gilbert, Grove Karl. "The Moon's Face: A Study of the Origin of Its Features." *Bulletin of the Philosophical Society of Washington,* 1892, Vol. 12, pp. 240-292.

Goguen, Jay D., and William M. Sinton. "Characterization of Io's Volcanic Activity by Infrared Polarimetry." *Science,* October 4, 1985.

Goldberg, Bruce A., Glenn W. Garneau, and Susan K. LaVoie. "Io's Sodium Cloud." *Science,* November 2, 1984.

Golden, Frederic. "Visit to a Large Planet." *Time,* November 24, 1980.

Goldman, Stuart J. "The Legacy of Phobos 2." *Sky & Telescope,* February 1990.

Gore, Rick. "Saturn: Riddles of the Rings." *National Geographic,* July 1981.

Hansen, C. J., et al. "Surface and Airborne Evidence for Plumes and Winds on Triton." *Science,* October 19, 1990.

Harris, Joel. "Bound for Jupiter." *Astronomy,* January 1990.

Hartmann, William K. "Birth of the Moon." *Natural History,* November 1989.

Ingersoll, Andrew P., and Kimberly A. Tryka. "Triton's Plumes: The Dust Devil Hypothesis." *Science,* October 19, 1990.

Johnson, Torrence V., et al.:
"Io: Evidence for Silicate Volcanism in 1986." *Science,* December 2, 1988.
"Volcanic Hotspots on Io: Stability and Longitudinal Distribution." *Science,* October 12, 1984.

Johnson, Torrence V., Robert Hamilton Brown, and Laurence A. Soderblom. "The Moons of Uranus." *Scientific American,* April 1987.

Johnson, Torrence V., and Laurence A. Soderblom. "Io." *Scientific American,* December 1983.

Kirk, Randolph L., Robert H. Brown, and Laurence A. Soderblom. "Subsurface Energy Storage and Transport for

Solar-Powered Geysers on Triton." *Science,* October 19, 1990.

Kolvoord, Robert A., Joseph A. Burns, and Mark R. Showalter. "Periodic Features in Saturn's F Ring: Evidence for Nearby Moonlets." *Nature,* June 21, 1990.

Kuiper, Gerard P. "On the Origin of the Irregular Satellites." *Proceedings of the National Academy of Sciences,* November 15, 1951.

Lewis, John S. "Satellites of the Outer Planets: Their Physical and Chemical Nature." *Icarus,* 1971, Vol. 15, pp. 174-185.

Lunine, Jonathan I. "Voyager at Triton." *Science,* October 19, 1990.

Nelson, Robert M., et al. "Temperature and Thermal Emissivity of the Surface of Neptune's Satellite Triton." *Science,* October 19, 1990.

"A New Moon for Saturn . . . and More to Come?" *Sky & Telescope,* November 1990.

Osterbrock, Donald E. "The Nature of Saturn's Rings: James E. Keeler's 'Prettiest Application of Doppler's Principle'." *Mercury,* March-April 1985.

Owen, Tobias. "Titan." *Scientific American,* February 1982.

Peale, Stanton. "Melting of Io by Tidal Dissipation." *Science,* March 2, 1979.

Pollack, James B., Joseph A. Burns, and Michael E. Tauber. "Gas Drag in Primordial Circumplanetary Envelopes: A Mechanism for Satellite Capture." *Icarus,* 1979, Vol. 37, pp. 587-611.

Pollack, James B., and Jeffrey N. Cuzzi. "Rings in the Solar System." *Scientific American,* November 1981.

Pollack, James B., Joel M. Schwartz, and Kathy Rages. "Scatterers in Triton's Atmosphere: Implications for the Seasonal Volatile Cycle." *Science,* October 19, 1990.

Robertson, Donald Frederick. "Cassini." *Astronomy,* September 1987.

Sagan, Carl. "Harold Clayton Urey: 1893-1981." *The Planetary Report,* July-August 1982.

Sagdeev, R. Z., and A. V. Zakharov. "Brief History of the Phobos Mission." *Nature,* October 19, 1989.

"Saturn's Rings Thinner Than Believed." *Astronomy,* November 1984.

Shoemaker, Eugene M. "The Geology of the Moon." *Scientific American,* December 1964.

Showalter, Mark R., et al. "Discovery of Jupiter's 'Gossamer' Ring." *Nature,* August 8, 1985.

Smith, Bradford A. "Voyage of the Century." *National Geographic,* August 1990.

Soderblom, Laurence A., et al. "Triton's Geyser-like Plumes: Discovery and Basic Characterization." *Science,* October 19, 1990.

Soderblom, Laurence A., and Torrence V. Johnson. "The Moons of Saturn." *Scientific American,* January 1982.

"Soviet Findings from Phobos and Mars." *Science News,* October 28, 1989.

Spaute, Dominique, and Richard Greenberg. "Collision Mechanics and the Structure of Planetary Ring Edges." *Icarus,* 1987, Vol. 70, pp. 289-302.

Squyres, Steven W., and Ray T. Reynolds. "The Solar System's Other Ocean." *The Planetary Report,* May-June 1983.

Strom, Robert G., Steven K. Croft, and Joseph M. Boyce. "The Impact Cratering Record on Triton." *Science,* October 19, 1990.

Taylor, R. L. S.:
"The Damocles Hypothesis." *Space,* March-April 1990.
"The Damocles Hypothesis: Part 2." *Space,* September-October 1990.

Terrile, Richard J. "Return to the Rings." *The Planetary Report,* October-November 1981.

Thompson, W. Reid, and Carl Sagan. "Color and Chemistry on Triton." *Science,* October 19, 1990.

Trauger, John T. "The Jovian Nebula: A Post-Voyager Perspective." *Science,* October 19, 1984.

Veverka, Joseph, Peter Thomas, and Thomas Duxbury. "The Puzzling Moons of Mars." *Sky & Telescope,* September 1978.

"Voyager's Last Picture Show." *Sky & Telescope,* November 1989.

Wegener, Alfred. "The Origin of Lunar Craters." *The Moon,* November-December 1975.

Wiesel, William. "Inelastic Collisions in Narrow Planetary Rings." *Icarus,* 1987, Vol. 71, pp. 78-90.

Zakharov, Aleksandr V. "Close Encounters with Phobos." *Sky & Telescope,* July 1988.

Other Sources

Ashworth, William B. "The Face of the Moon: Galileo to Apollo." Exhibition catalog. Kansas City, Mo.: Linda Hall Library, 1989.

National Aeronautics and Space Administration:
Photograph caption, *Voyager 2.* Photo number P23936/S-2-32. Greenbelt, Md.: Goddard Space Flight Center, no date.
Photograph caption, *Voyager 2.* Photo number P24065/S-2-66. Greenbelt, Md.: Goddard Space Flight Center, no date.
Photograph caption, *Voyager 2.* Photo number P24067/S-2-68. Greenbelt, Md.: Goddard Space Flight Center, no date.
Photograph caption, *Voyager 2.* Photo number P24308/S-2-78. Greenbelt, Md.: Goddard Space Flight Center, no date.
Photograph caption, *Voyager 2.* Photo number P29481/U-2-14. Pasadena, Calif.: Jet Propulsion Laboratory, no date.

"New Moon Discovered Orbiting Planet Saturn." *NASA News,* press release. Washington, D.C.: NASA, July 24, 1990.

Reiber, Duke B. (ed.). Proceedings of the NASA Mars Conference. Volume 71, Science and Technology Series. San Diego, Calif.: Univelt, 1988.

"To Uranus and Beyond." NASA publication no. 400-303. Pasadena, Calif.: Jet Propulsion Laboratory, January 1987.

"The Voyager Flights to Jupiter and Saturn." NASA publication no. JPL 400-148. Pasadena, Calif.: Jet Propulsion Laboratory, July 1982.

INDEX

lution, 71, 78, 86, 88; Galileo mission to, 120; heating of, 72, 73, 74, 78, 80, 86, 88; Kuiper's ideas about, 69-71; in light-speed measurement, 67; movements, *65;* naming of, by Marius, 66; orbits, *84, 86, 88;* size, 71, 86, 88; surface features of Callisto, *66-67, 76-77,* 78, 79, *86-87;* surface features of Europa, *51, 68,* 73, 76, 77-78, *88-89;* surface features of Ganymede, *67,* 73, 76, 78, 79, *86-87;* surface features of Io, *51, 69,* 72, *74-75,* 76, 77, *88-89;* Voyager flybys, 74, 76-78, 88

Galileo Galilei, 9; crater named for, *16;* Jovian moons discovered by, record of, *64,* 65-66; lunar studies, 14, *15;* quoted, 14, 65, 66, 95; rings of Saturn viewed by, 95

Galileo mission, 120

Ganymede (Jovian moon): density and composition, 71, 74, 86; evolution, 78, 86; orbit, *84, 86;* size, 71, 86; surface effects of heating, 73, 79-80, 86; surface features, *67,* 76, 78, 79, *86-87;* Voyager views, 76, 78

Genesis Rock (lunar sample), *39*

Geysers, nitrogen: Triton, 116-*117*

Giant impact theory of Moon's origin, 33, *36,* 38, 40, 42, 48

Gilbert, Grove Karl, theory propounded by, 19-21; evidence supporting, 22

Gold, Thomas, 31-32

Gossamer ring: Jupiter, *79, 90-91*

Gravitational interactions: in ring systems, 97, 99, 106, *107,* 109, *124-127,* 132. *See also* Tidal forces

Grimaldi, Francesco, 16

Gruithuisen, Franz von Paula, 19

H

Hadley Rille, Moon, *30*

Hale, George Ellery: Keeler's letter to, *96*

Hall, Asaph, 52-*54*

Halo (ring around Jupiter), *79, 90-91*

Hansen, Candice, 98

Hartmann, William, 38, 40

Heating: of Galilean satellites, 72, 73, 74, 78, 80, 86, 88; tidal flexing and, *72-73,* 74, 78, 88

Helene (Saturnian moon), *110*

Herschel, William, 18, 52

Herschel crater, Mimas, 102

Hevelius (Hewelcke), Johannes, 16

Himalia (Jovian moon): discovery, 68; as prograde irregular, 70, *85*

Hooke, Robert, 18

Horstman, Kevin C., 58

Hubbard, William, 99

Hubble Space Telescope, images by: Pluto and Charon, *118;* Saturn and rings, *128-129*

Hunt, Garry: quoted, 78

Hunten, Donald, 109, 110

Huygens, Christiaan, 95, 108

Hyperion (Saturnian moon), *10,* 95, *111*

I

Iapetus (Saturnian moon), 102, *111*

Ice on Galilean satellites, *66, 67, 68,* 79, 86

Ice volcanoes, possible: Ariel, 112; Enceladus, 108

Ilmenite in Moon rock, *39*

Impact basins on Moon, *29,* 44; mare formed in, *46*

Impact craters. *See* Craters, lunar

Impact theories: giant impact theory, 33, *36,* 38, 40, 42, 48; of lunar cratering, 18-19, 20-22; of Miranda, 113-114, *115*

Io (Jovian moon), *51,* 66, *69,* 74; clouds around, 74, *76;* density, 71, 88; evolution, 78, 88; in light-speed measurement, 67; orbit, *84,* 88; tidal heating, 72, 74, 88; volcanoes, *69,* 72, *74-75,* 77, 80, *88-89;* Voyager views of, *75,* 76, *88-89*

Irregular moons, 70; capture scenario for, *34-35,* 60, 70; circularization of orbits, *60-61;* of Jupiter, retrograde vs. prograde, 69, 70, *84-85;* of Neptune, orbits of, 115, *116;* of Saturn, 101, 102, *108*

J

Janus (Saturnian moon), *109*

Jupiter (planet), moons of, *10,* 54, *80-81;* Amalthea, *Voyager 1* views of, *71,* 77; capture hypothesis, 25, 70; discovery and naming of, 64, 65-66, 67, 68-69, 70, 80; irregulars, retrograde vs. prograde, 69, 70, *84-85;* orbits, 70, *84-85, 86, 88, 90-91;* origin, 69-70, 84. *See also* Galilean satellites

Jupiter (planet), rings of, 77, *79,* 80, 84, *90-91,* 98; composition, 99

K

Keeler, James, experiment by, 96-97; letter describing, *97*

Kepler, Johannes: crater named for, *16;* lunar studies, 14-15; Martian moons predicted by, 9, 52, 54

Kowal, Charles, 70

Kuiper, Gerard: Galilean satellites, ideas about, 69-71; moons classified by, 69-70; moons discovered by, *11,* 115; Titan's atmosphere proved by, 108

L

Leda (Jovian moon), 70; orbit, *85*

Lewis, John, 74

Line of nodes, *13*

Loki (volcano), Io, *69,* 74-*75,* 77

Luna missions, 32; *Luna 3,* far side of Moon viewed by, 32, 48; *Luna 9,* 32; *Luna 24,* 32

Lunar origin theories: Earth's moon, 22-25, *24,* 28, 33, *36,* 37-38, 40, 42, 48; fission hypothesis, 23-25, *24,* 36, 37; giant impact theory, 33, *36,* 38, 40, 42, 48; Jovian moons, 69-70, 78, 84; in Kuiper's classification system, 69-70; Martian moons, 59, 60, 64, 82; two-moon, 48. *See also* Accretion theory of moon formation; Capture hypothesis of lunar origins

Lysithea (Jovian moon): discovery, 68; as prograde irregular, 70, *85*

M

Mädler, Johann Heinrich von, 19; lunar map by, *17*

Magma: Moon, 39, *42-43,* 46

Magnetic field lines and ring spokes, theory of, 122, *123*

Main asteroid belt, *81*

Mapping: of Moon, earliest, 14, *15-17;* of Phobos, 58

Mare Crisium, Moon: features of, *17*

Mare Imbrium, Moon: engraving of, *15;* layers of, 30; rille, *30*

Mare Nectaris, Moon: crater by, *29*

Mare Nubium (Mare Cognitum), Moon, 32

Mare Orientale, Moon, *29*

Maria, Moon's, 37, 41, *46-47;* near and far sides, difference in, *48-49;* origin of, *43;* rock sample from, *39;* Urey's theory, 29-30

Mariner spacecraft, 56; *Mariner 7,* 56; *Mariner 9,* 56-57, 58

Mars (planet), moons of, *10, 56-57, 82-83;* as captured asteroids, 59, 60, 64, 82; craters, *56, 57,* 58, 59, *82;* darkness, 58, 59; discoverer (Asaph Hall), 52-*54;* Kepler's prediction of, 9, 52, 54; in Kuiper's system, 70; orbital destinies, opposite, 64, 82; orbits, 53, 54, 55, 59, 60, 63, *82-83;* probes investigating, 51, 56-57, 58, 59, 64-65; regolith, 57-58, 59, 83. *See also* Phobos

Mayer, Johann Tobias: lunar mapping by, *16,* 17

Meteorite craters: Earth, 21, 30

Meteorite theory of lunar-crater origin, 18-19, 20-22, *44-45, 47*

Methane in Titan's atmosphere, 108; relative amount of, 109, 110

Metis (Jovian moon), 80; orbit, *84,*

ACKNOWLEDGMENTS

The editors wish to thank James Arnold, University of California—San Diego, La Jolla; Stephen Brush, University of Maryland, College Park; Steven J. Dick, U.S. Naval Observatory, Washington, D.C.; James B. Garvin, Goddard Space Flight Center, Greenbelt, Md.; William Hartmann, Planetary Science Institute, Tucson, Ariz.; Kathy Hoyt, U.S. Geological Survey, Flagstaff, Ariz.; Richard Kerr, Science/American Association for the Advancement of Science, Washington, D.C.; Heidi Klein, Bildarchiv Preussischer Kulturbesitz, Berlin, Germany; Alfred McEwen, U.S. Geological Survey, Flagstaff, Ariz.; H. J. Melosh, University of Arizona, Tucson, Ariz.; Scott Murchie, Brown University, Providence, R.I.; Jacques-Clair Noëns, Observatoire du Pic-du-Midi, France; David Pieri, Jet Propulsion Laboratory, Pasadena, Calif.; Carolyn Porco, University of Arizona, Tucson, Ariz.; Peter Schultz, Brown University, Providence, R.I.; Mark Showalter, NASA Ames Research Center, Moffett Field, Calif.; G. Jeffrey Taylor, Sinclair Library, Honolulu, Hawaii; Ray Villard, Space Telescope Science Institute, Baltimore, Md.; Marie-Josée Vin, Observatoire de Haute-Provence, France; John Wood, Smithsonian Astrophysical Observatory, Cambridge, Mass.

PICTURE CREDITS

The sources for the illustrations in this book are listed below. Credits from left to right are separated by semicolons, from top to bottom by dashes.

Cover: Art by Don Davis. 6, 7: NASA, Johnson Space Flight Center. 8: Initial cap, detail from pages 6, 7. 10, 11: Art by Matt McMullen. 12, 13: Art by Matt McMullen (2)—Lick Observatory, courtesy Pat Shand (7). 15, 16: Linda Hall Library. 17: Bildarchiv Preussischer Kulturbesitz, Berlin; Linda Hall Library. 20: Art by Time-Life Books—Peter L. Kresan; NASA, Johnson Space Flight Center. 24: Bettmann Archive—Larry Sherer from *Journal of Geophysical Research,* Vol. 68, No. 5. 26, 27: Art by Matt McMullen. 28: NASA/Lunar Orbiter. 29: NASA, Johnson Space Flight Center; NASA/Lunar Orbiter (2). 30, 31: NASA/Lunar Orbiter. 33-36: Art by Yvonne Gensurowsky. 39: NASA, except middle right, Russell Chappell/NASA. 41: Art by Don Davis, courtesy Paul Spudis. 42, 43: Art by Don Davis. 44, 45: Art by Don Davis, courtesy Paul Spudis; NASA, Johnson Space Flight Center, inset art by Don Davis. 46, 47: Art by Don Davis, courtesy Paul Spudis; NASA, Johnson Space Flight Center, inset art by Don Davis. 48, 49: NASA, Johnson Space Flight Center; Lick Observatory, Calif. 50: G. Avenasov and B. Zhukov, IKI, USSR, color processing by Mark Robinson and Fraser Fanale, Planetary Geosciences Division, University of Hawaii, in collaboration with the Soviet Phobos Imaging Team. 51: NASA. 52: Initial cap, detail from page 51. 54, 55: U.S. Naval Observatory. 56, 57: NASA. 60-63: Art by Alfred T. Kamajian. 64: Emmett Bright, Rome, courtesy Biblioteca Nazionale Centrale, Florence. 65: Dennis di Cicco. 66, 67: NASA/JPL. 68: NASA/JPL; Alfred S. McEwen/U.S. Geological Survey, Flagstaff, Ariz. 69: U.S. Geological Survey, Flagstaff, Ariz.; NASA/JPL. 71: NASA/JPL. 72, 73: Art by Alfred T. Kamajian. 74: Art by Alfred T. Kamajian. 75: NASA/JPL. 76: Bruce A. Goldberg, Glenn W. Garneau, and Susan K. LaVoie—JPL (California Institute of Technology). 79: NASA, Ames Research Center, courtesy Mark Showalter. 80, 81: Art by Stephen R. Wagner. 82, 83: Art by Stephen R. Wagner; insets NASA, courtesy Joseph Veverka. 84, 85: Art by Stephen R. Wagner. 86, 87: Art by Stephen R. Wagner; NASA, courtesy Joseph Veverka (2), art by Stephen R. Wagner (2). 88, 89: Art by Stephen R. Wagner; U.S. Geological Survey, Flagstaff, Ariz. (2), art by Steven R. Wagner (2). 90, 91: Art by Stephen R. Wagner (2), inset NASA/JPL. 92, 93: NASA/JPL. 94: Initial cap, detail from pages 92, 93. 96, 97: Larry Sherer, courtesy U.S. Naval Observatory, Washington, D.C.—Yerkes Observatory, Williams Bay, Wis. 100: NASA/U.S. Geological Survey, Flagstaff, Ariz.—NASA, courtesy National Space Science Data Center—NASA/JPL. 102-107: Art by Matt McMullen. 108: Art by Fred Holz. 109: NASA/Ames Research Center; NASA, courtesy National Space Science Data Center (5); NASA/U.S. Geological Survey, Flagstaff, Ariz. 110, 111: NASA, courtesy National Space Science Data Center; NASA/U.S. Geological Survey, Flagstaff, Ariz.; NASA, courtesy National Space Science Data Center (2); NASA/U.S. Geological Survey, Flagstaff, Ariz.; NASA, courtesy National Space Science Data Center; NASA/JPL; NASA, courtesy National Space Science Data Center (2); NASA/U.S. Geological Survey, Flagstaff, Ariz.; NASA, courtesy National Space Science Data Center. 112: NASA/JPL. 114, 115: Art by Fred Holz (2); W. M. Sinton, Institute for Astronomy, University of Hawaii; U.S. Geological Survey, Flagstaff, Ariz.; NASA/JPL. 116, 117: Art by Fred Holz (2); Larry Sherer from *Sky & Telescope,* Feb. 1990, p. 155; NASA/U.S. Geological Survey, Flagstaff, Ariz. (2). 118: Art by Fred Holz—NASA/ESA Space Telescope Science Institute. 120, 121: Art by Yvonne Gensurowsky. 122, 123: Art by Yvonne Gensurowsky, except ring/spoke blowup art, by Matt McMullen. 124-127: Art by Yvonne Gensurowsky. 128, 129: NASA/ESA Space Telescope Institute, except F ring and ring system art, by Yvonne Gensurowsky. 130, 131: Anglo-Australian Telescope Board, inset art by Yvonne Gensurowsky. 132: NASA/JPL (2), inset art by Time-Life Books.

Time-Life Books is a division of Time Life Inc.,
a wholly owned subsidiary of
THE TIME INC. BOOK COMPANY

TIME-LIFE BOOKS

Managing Editor: Thomas H. Flaherty
Director of Editorial Resources:
Elise D. Ritter-Clough
Director of Photography and Research:
John Conrad Weiser
Editorial Board: Dale M. Brown, Roberta Conlan,
Laura Foreman, Lee Hassig, Jim Hicks, Blaine
Marshall, Rita Thievon Mullin, Henry Woodhead

PUBLISHER: Joseph J. Ward

Associate Publisher: Trevor Lunn
Editorial Director: Donia Ann Steele
Marketing Director: Regina Hall
Director of Design: Louis Klein
Production Manager: Prudence G. Harris
Supervisor of Quality Control: James King

Editorial Operations
Production: Celia Beattie
Library: Louise D. Forstall
Computer Composition: Deborah G. Tait
(Manager), Monika D. Thayer, Janet Barnes
Syring, Lillian Daniels

Correspondents: Elisabeth Kraemer-Singh (Bonn),
Christine Hinze (London), Christina Lieberman
(New York), Maria Vincenza Aloisi (Paris), Ann
Natanson (Rome). Valuable assistance was also
provided by Elizabeth Brown (New York), Judy
Aspinall (London).

VOYAGE THROUGH THE UNIVERSE

SERIES EDITOR: Roberta Conlan
Series Administrator: Norma E. Shaw

Editorial Staff for *Moons and Rings*
Art Director: Cynthia T. Richardson
Picture Editor: Tina McDowell
Text Editors: Stephen Hyslop,
Robert M. S. Somerville
Associate Editor/Research: Mary H. McCarthy
Assistant Editors/Research: Patricia A. Mitchell,
Quentin Story, Elizabeth Thompson
Writers: Mark Galan, Darcie Conner Johnston
Assistant Art Director: Brook Mowrey
Editorial Assistant: Katie Mahaffey
Copy Coordinator: Juli Duncan
Picture Coordinator: David Beard

Special Contributors: J. Kelly Beatty, Robert
Cane, Andrew Chaikin, George Constable, James
Dawson, Gina Maranto, Eliot Marshall, Mark
Washburn (text); Jocelyn G. Lindsay, Cheryl Pel-
lerin, Eugenia Scharf (research); Barbara L. Klein
(index).

CONSULTANTS

JOSEPH A. BURNS, professor and chairman of theoretical and applied mechanics at Cornell University in Ithaca, New York, is also a professor of astronomy. His research concerns planetary rings,
interplanetary dust, satellites, and comets.

A. G. W. CAMERON, a professor of astronomy at
Harvard College Observatory in Cambridge, Massachusetts, conducts theoretical research in astrophysics and planetary sciences.

JEFFREY N. CUZZI is a research scientist at NASA's
Ames Research Center, Moffett Field, California,
where he studies planetary ring structure and dynamics, solar system formation processes, and the
structure of cometary nuclei. He is a member of the
Voyager Scientific Imaging Team and is an interdisciplinary scientist for rings and dust with the
Cassini Mission.

RONALD E. DOEL is a postdoctoral historian at the
Center for History of Physics of the American Institute of Physics in New York City. His areas of
research include the history of modern astronomy
and geophysics.

ALAN W. HARRIS is a supervisor at the Earth and
Planetary Physics Group at the Jet Propulsion Laboratory in Pasadena, California, where his interests
include orbital dynamics, the origin and evolution
of planets, satellites, and rings, and the study of
asteroids.

JONATHAN I. LUNINE, an associate professor of
planetary science and theoretical astrophysics at
the University of Arizona in Tucson, is interested in
the interiors, surfaces, and atmospheres of outer
solar system bodies and low-mass stars. He was a
guest associate of the Voyager Ultraviolet Spectrometer team and is an interdisciplinary scientist
on the Cassini Mission.

WILLIAM B. MCKINNON is an associate professor
of earth and planetary sciences at Washington University in Saint Louis, Missouri. His research fo-
cuses on impact cratering, the satellites of the outer
planets, and Pluto.

PAUL D. SPUDIS, a senior staff scientist at the Lunar
and Planetary Institute, Houston, Texas, is a principal investigator in the NASA Planetary Geology
Program. His research includes geology and petrol-
ogy of the lunar crust.

JOSEPH VEVERKA is a professor of astronomy and
planetary science at Cornell University in Ithaca,
New York, and a member of the Voyager, Galileo,
and Mars Observer Imaging teams. His recent work
emphasizes studies of the present state and past
evolution of surfaces of the Moon.

**Library of Congress Cataloging in
Publication Data**
Moons and rings / by the editors of Time-Life
Books.
p.   cm.   (Voyage through the universe).
Bibliography: p.
Includes index.
ISBN 0-8094-8450-1.
ISBN 0-8094-8451-X (lib. bdg.).
1. Moon. 2. Satellites.
3. Astronautics in astronomy.
I. Time-Life Books. II. Series.
QB581.M66   1991
523.2—dc20   90-47599 CIP

For information on and a full description of
any of the Time-Life Books series, please call
1-800-621-7026 or write:
Reader Information
Time-Life Customer Service
P.O. Box C-32068
Richmond, Virginia 23261-2068

Earth: diameter 7,926 miles

Neptune: diameter 30,775 miles

Uranus: diameter 31,763 miles

Red supergiant: diameter 400 million miles

Solar System: diameter 7.5 billion miles

Globular cluster: diameter 2×10^{14} miles

Milky Way: diameter 100,000 light-years

Local Group of galaxies:
6 million light-years across

Largest double radio source:
length 17 million light-years